Holt Algebra 2

Technology Lab Activities

HOLT, RINEHART AND WINSTON
A Harcourt Education Company
Orlando • Austin • New York • San Diego • London

Printed in the United States of America

ISBN 0-03-078456-5

1 2 3 4 5 6 7 8 9 862 09 08 07 06

Contents

Name ______________________________ Date __________ Class __________

LESSON 1-5

Technology Lab Recording Sheet (p. 33)

Explore Negative Exponents

Try This

Activity

Use a table to evaluate each pair of expressions for the given *x*-values.

1. 2^x and 2^{-x} for $x = -2, -1, 0, 1,$ and 2
   Complete the table.

x	2^x	2^{-x}
–2		
–1		
0		
1		
2		

2. 3^x and 3^{-x} for $x = -4, -3, -2, -1,$ and 0
   Complete the table.

x	3^x	3^{-x}
–4		
–3		
–2		
–1		
0		

3. 4^x and 4^{-x} for $x = -2, -1, 0, 1,$ and 2
   Complete the table.

x	4^x	4^{-x}
–2		
–1		
0		
1		
2		

Name ______________________ Date ___________ Class ___________

LESSON 1-5

Technology Lab

Explore Negative Exponents continued

4. 5^x and 5^{-x} for $x = -4, -2, 0, 2,$ and 4
Complete the table.

x	5^x	5^{-x}
–4		
–2		
0		
2		
4		

5. Use a table to compute 2^{-x} and $\frac{1}{2^x}$ for $x = 0, 1, 2, 3, 4,$ and 5.
Complete the table.

x	2^x	$\frac{1}{2^x}$
–2		
–1		
0		
1		
2		

Explain what you notice. ____

____

____

6. Make a Conjecture Use the tables you created for Problems 1–5 to calculate the product of Y1 and Y2 for each value of x. Then make a conjecture about the product of a^x and a^{-x} for any value of a.

____

____

____

7. Make a Conjecture For what values of x is $2^{-x} > 2^x$?

____

Name ______________________ Date ____________ Class ____________

LESSON **1-9**

Technology Lab
Introduction to Parent Functions

Use with Lesson 1-9

Activity

Use your calculator to determine how $y = x^2 + 12$ is transformed from its parent function $y = x^2$.

Step 1 Use your graphing calculator to graph the parent function of $y = x^2$.
Enter the following key sequence:

Press [Y=] [X,T,θ,n] [x^2] [GRAPH].

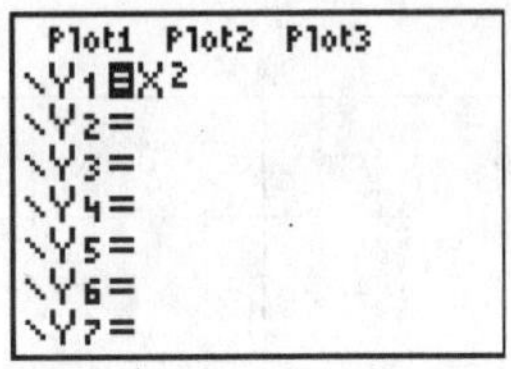

The graph of $y = x^2$ shows a parabola.

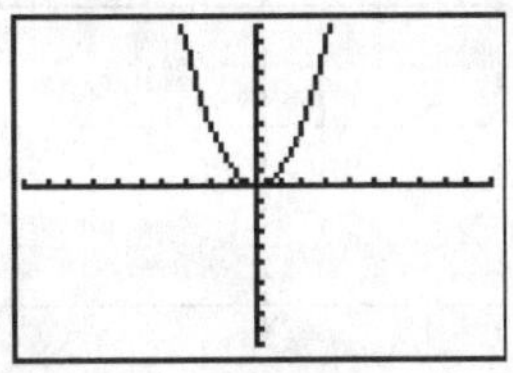

Step 2 Now graph $y = x^2 + 12$ on the same graph.

Enter the following key sequence:

Press [Y=].

Use your arrow key and scroll down to Y2.

Press [ENTER]. The equal sign should be highlighted.

Then press the right arrow key one time: [▶]

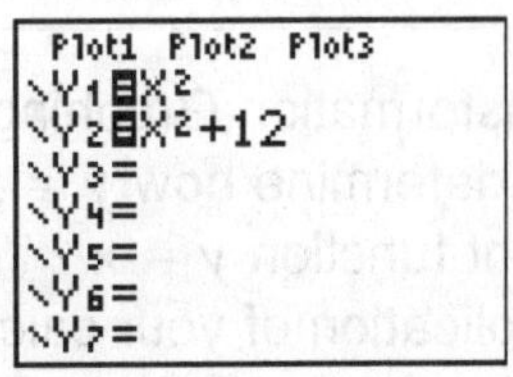

Now press [X,T,θ,n] [x^2] [+] 12 [GRAPH].

When you look at the screen, the graph of $y = x^2 + 12$ is not visible so you must change the window settings. In this case, the maximum visible range of the window, **Ymax=**, needs to be changed.

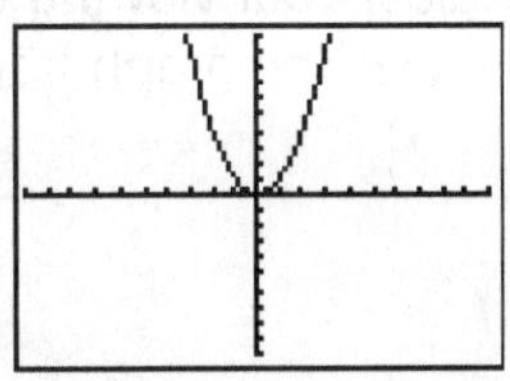

Enter the following key sequence to change the **Ymax=** setting:

Press [WINDOW].

Use your arrow key to scroll down to **Ymax=**.
Change the **Ymax=** to 20.

Press [GRAPH].

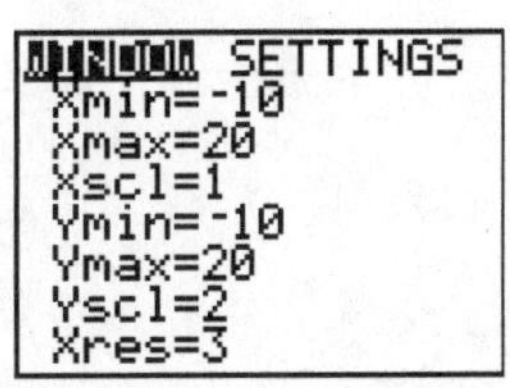

Name ______________________ Date ___________ Class ___________

LESSON 1-9

Technology Lab

Introduction to Parent Functions continued

Now you can see both graphs. Not that the function $y = x^2 + 12$ is translated 12 units up from the parent function $y = x^2$.

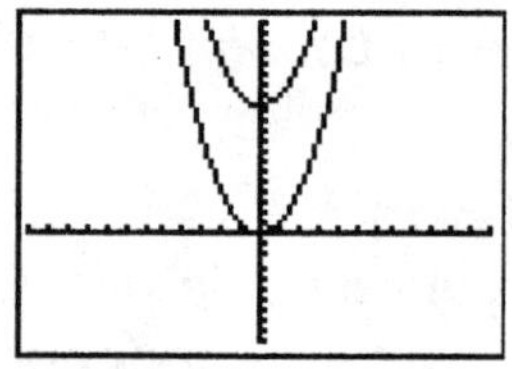

Try This

1. Use your graphing calculator and graph $y = (x - 2)^2$. How does this function transformed from its parent function $y = x^2$?

Graph:

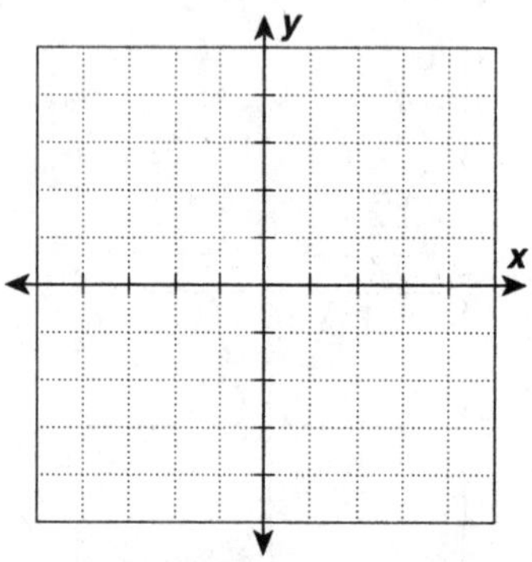

____

2. **Critical Thinking:** What is the new function if $y = x^2 + 12$ is transformed 4 units down?

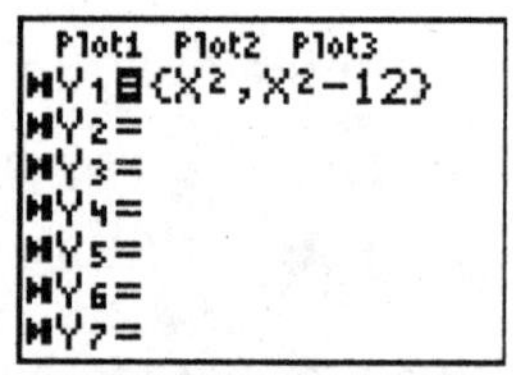

____

3. Use the Transformation Graphing application on your calculator to determine how $y = x^2 - 4$ is transformed from its parent function $y = x^2$. To use the Transformation Graphing application of your calculator, press APPS ALPHA 4 and use ▼ to move the cursor to **Transfrm**. Then press ENTER to begin the application. Now press Y= and enter the functions as a sequence. Adjust your viewing window and press GRAPH. How is the graph transformed?

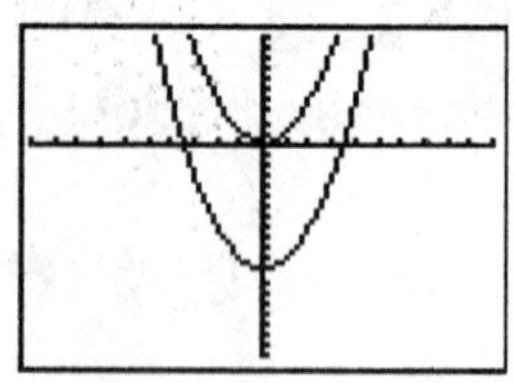

____

Name ______________________________ Date ____________ Class ____________

LESSON 2-3

Technology Lab Recording Sheet (pp. 113–114)

Explore Graphs and Windows

Try This

Activity 1

Graph each function in a square window that shows both the *x*- and *y*-intercepts.

1. $y = 2x - 25$

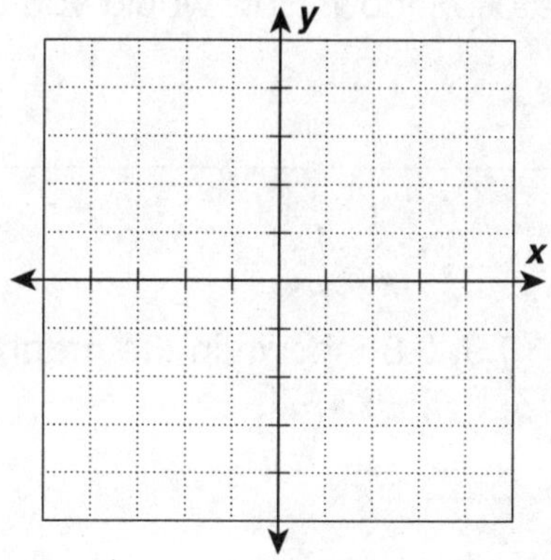

2. $y = -3x - 50$

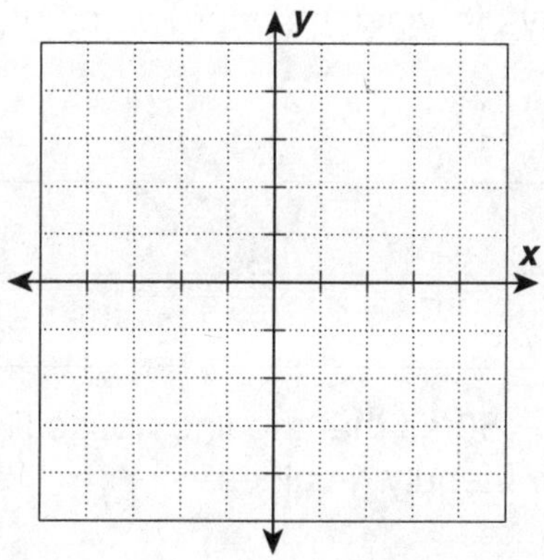

3. $y = 20 + 0.8x$

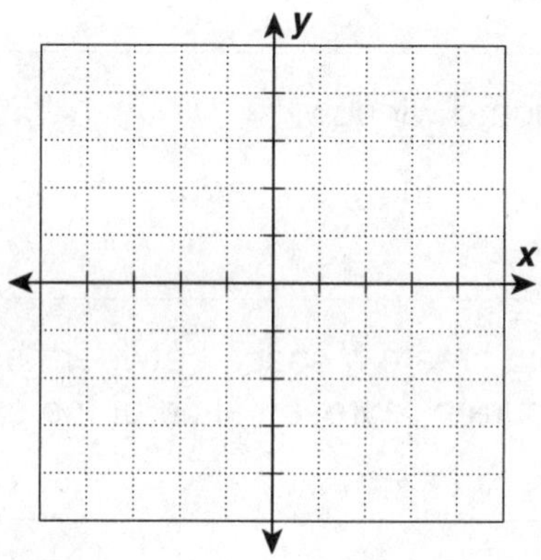

4. When you enter and graph a function and only a piece of the graph is visible in the lower left, what adjustments can you make to see the key features of the graph?

____

Name ______________________ Date __________ Class __________

LESSON 2-3

Technology Lab

Explore Graphs and Windows continued

5. How can you see the graph of $y = 0.01x$?

____

6. **What if...?** Suppose that you wanted to make $y = 0.5x$ look very steep or $y = 10x$ look flat in the calculator window. How would you change the window settings?

____

Activity 2

1. If you ZOOM out on the point (0.9, 0.8) shown in the graph in part a, the window changes to [−8.5, 10.3] by [−5.4, 7].

   Find **Xmax – Xmin**. ____________

   Find **Ymax – Ymin**. ____________

   Use TRACE and the arrow keys to view the coordinates of points on the line. Why is the window friendly?

   ____

   ____

   ____

2. Explain how to graph $y = 3x - 50$ so that the line "looks like" a line with a slope of 3 and allows you to trace to friendly x-values.

   ____

   ____

3. Graph $y = x + 0.5$ in the standard window.

   a. How can you make the slope of the graph appear to be 1?

   ____

   b. Which ZOOM window would create a space between the y-intercept and the origin while keeping an accurate representation of the slope?

   ____

Name ______________________ Date __________ Class __________

LESSON **2-7**

Technology Lab

Graphing Your Motion

Graphs made using a Motion Detector can be used to study motion. In this experiment, you will use a Motion Detector to make graphs of your own motion.

OBJECTIVES

In this experiment, you will use a Motion Detector to measure distance and velocity; produce graphs of your motion; analyze the graphs you produce.

MATERIALS

TI-84 Plus graphing calculator; EasyData application
CBR 2 motion and direct calculator cable
or Motion Detector and data-collection interface
masking tape; meter stick

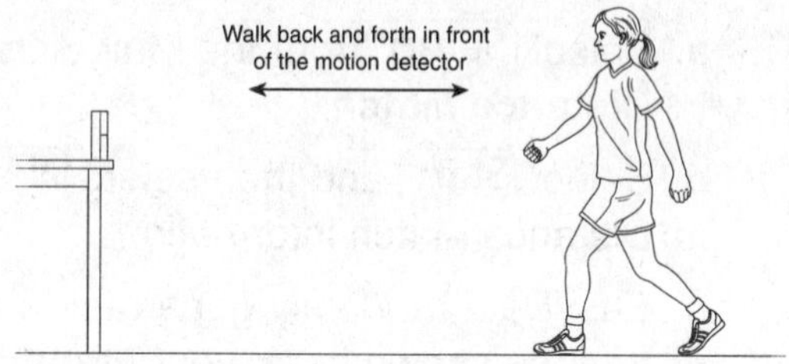

Activity 1 Distance *vs.* Time Graphs

1. Fasten a Motion Detector to a tabletop facing an area free of furniture and other objects. The Motion Detector should be at a height of about 15 cm above your waist level.
2. Use short strips of masking tape on the floor to mark the 1 m, 2 m, 3 m, and 4 m distances from the Motion Detector.
3. Connect the Motion Detector.
   - **a.** Open the pivoting head of the Motion Dectector.
   - **b.** If the Motion Detector has a sensitivity switch, set it to Normal.
   - **c.** Turn on the calculator and connect it to the Motion Detector. (This may require the use of a data-collection interface.)
4. Set up EasyData for data collection.
   - **a.** Start the EasyData application, if it is not already running.
   - **b.** Select File from the Main screen, and then select **New** to reset the application.
5. Explore making distance *vs.* time graphs.
   - **a.** Stand at the 1.0 m mark, facing away from the Motion Detector.
   - **b.** Signal your partner to select Start.
   - **c.** Slowly walk to the 2.5 m mark and stop.
   - **d.** When data collection ends, a graph of distance versus time will be displayed.
   - **e.** Sketch your graph on the empty graph provided at the right.
   - **f.** Select Main to return to the Main screen.

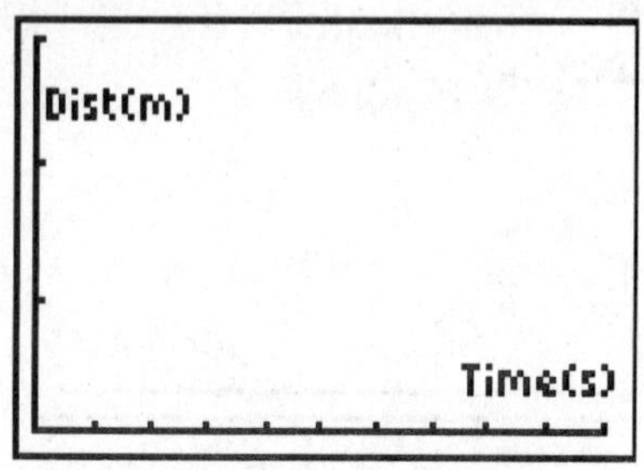

6. Repeat Step 5 while walking faster. Sketch your new line on the same graph.

Name ______________________ Date ____________ Class ____________

LESSON **2-7**

Technology Lab

Graphing Your Motion continued

Try This

1. Describe the differences between your two graphs. ______________________

____

Activity 2 Matching Distance *vs.* Time Graphs

7. Set up the calculator and interface for distance *vs.* time graph matching.
   a. Select (Setup) from the Main screen, then select **Distance Match**.
   b. Select (Start) and then select (Next) after reading the Distance Match information.
   c. EasyData randomly generates matching graphs like the one shown here; your graph may be different.

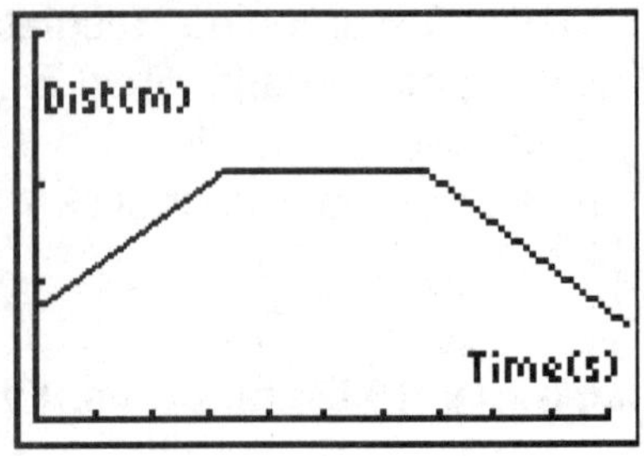

8. Match the first distance *vs.* time graph.
   a. Examine the graph and plan what you will do to match it. Note: The vertical axis runs from 0 to 3 meters. Data will be collected for 10 seconds.
   b. Take your starting position in front of the Motion Detector.
   c. When you are ready to begin matching, signal your partner to select (Start).
   d. Move according to your plan.
   e. Examine the graph of the results.
   f. Sketch your results on Graph 1 below. Describe what you had to do to match the first graph. Sketch the graph you were matching and the graph of your motion.
   g. If you want to repeat the first match, select (Retry). If you are ready to move on to a second graph, select (New).
9. Repeat Step 8 until you have matched a total of three distance *vs.* time graphs.

Try This

Sketch your results in the empty graphs provided below. Describe what you had to do to match each of the graphs.

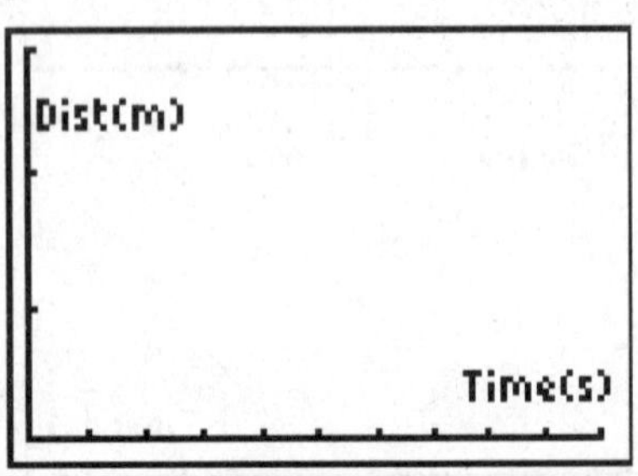

Graph 1

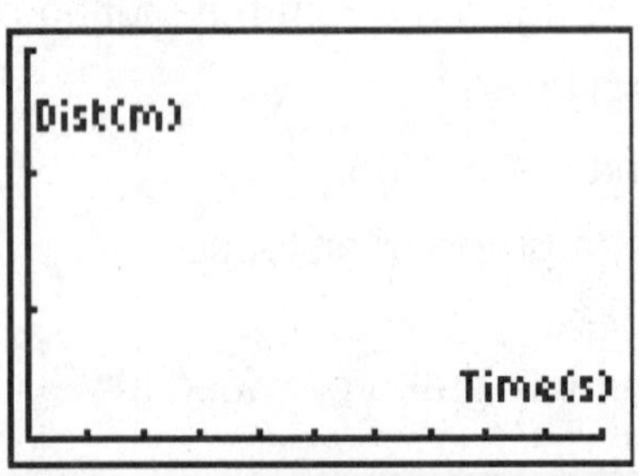

Graph 2

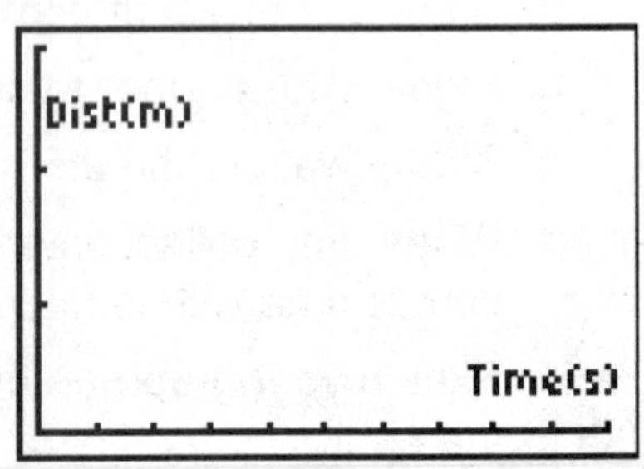

Graph 3

______________ ______________ ______________

______________ ______________ ______________

Name ______________________ Date ___________ Class ___________

LESSON 2-9

Technology Lab Recording Sheet (p. 157)

Solving Absolute-Value Equations

Try This

1. Solve $3|x - 1| = 6$ by using a table of values. Then solve by graphing.

____

x	y

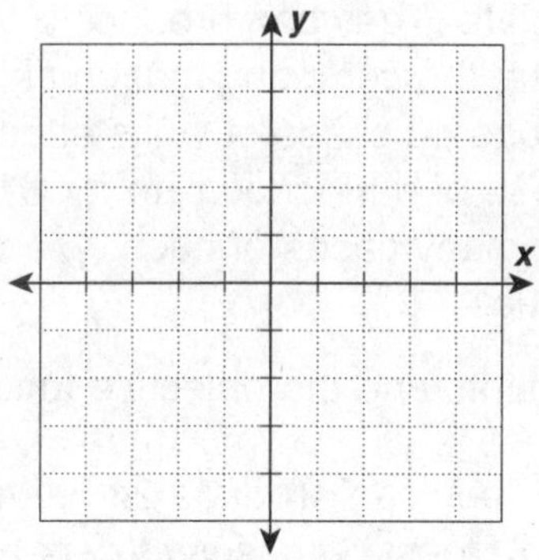

2. Solve $5|x + 3| = 0$ by using a table of values. Then solve by graphing.

____

X	Y1

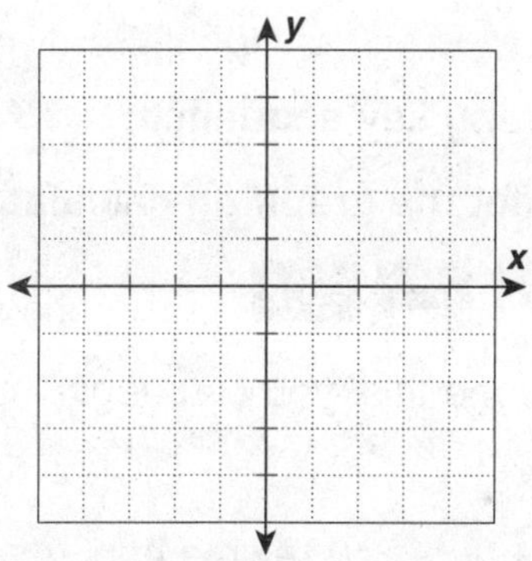

3. What happens when you solve $2|x + 1| = -4$ by using a table of values?

____

What happens when you solve by graphing?

____

X	Y1

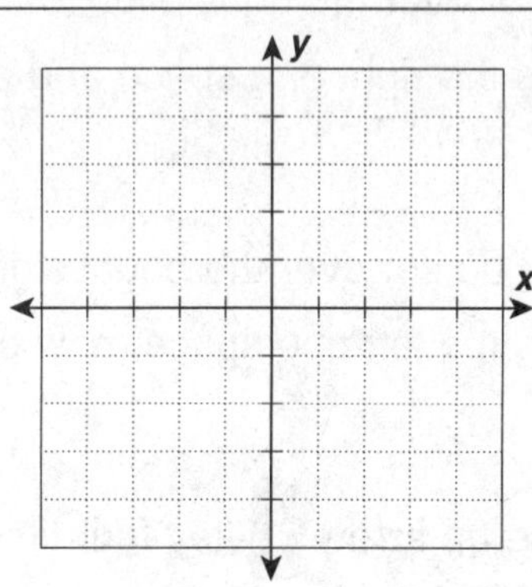

Name ______________________ Date ____________ Class ____________

LESSON **3-4**

Technology Lab

Linear Programing

Use with Lesson 3-4

Activity

The Swiss Chocolate Company makes two types of chocolates, dark chocolate and white chocolate. They can produce a total of 240 cases of chocolates. Based on previous sales, the company must make at least 60 cases of dark chocolate. They must produce 40 cases of white chocolate but no more than 160 cases. They can sell each case of dark chocolate for \$12.50 and each case of white chocolate for \$17.50. How many cases of each type of chocolate should they produce to maximize revenue?

Write the constraints and the objective function.

$x + y \leq 240$ They can produce 240 total cases;
$x \geq 60$ At least 60 cases of dark chocolate;
$y \geq 40$ They must also produce at least 40 cases of white chocolate;
$y \leq 60$ But no more than 160 cases of white chocolate.
$P = 12.5x + 17.5y$

Enter the following key sequence:

You must first set the graphing calculator for inequalities.

Press [APPS] [ALPHA] [x^2] [ENTER].

After you have done this, you can begin inputting the constraints.

1) To input $x + y \leq 240$ rewrite it as $y \leq 240 - x$.

Now Press [ALPHA] [≤]. Then press the Right arrow key [→] one time. Now enter **240** [−] [X,T,θ,n] [ENTER].

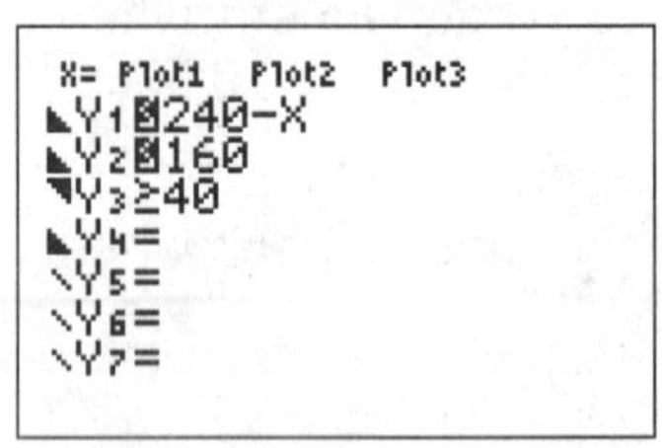

2) $x \leq 160$

Move the cursor over the equal sign and press [ALPHA] [≤]. Move to the right one space and enter 160 [ENTER].

3) $y \geq 40$

Now move the cursor over the equal sign and press [ALPHA] [≥]. Move to the left one space: now enter **40**.

4) $x \geq 60$

Then press the up arrow [↑] key four times. Now Press [ENTER] [ALPHA] [≥].

Move to the right one space and enter 60.

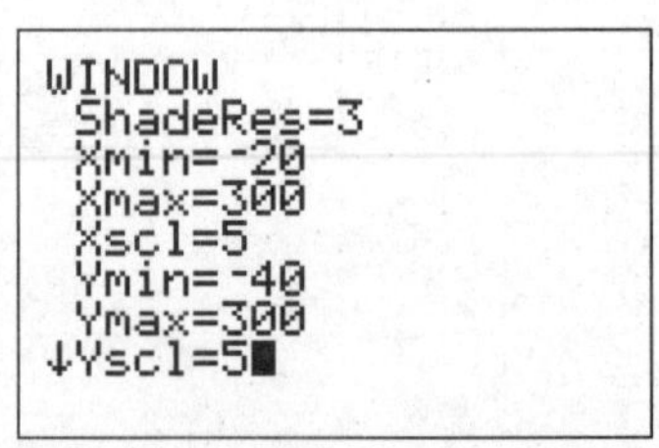

Name ______________________ Date ____________ Class ____________

LESSON 3-4

Technology Lab

Linear Programing continued

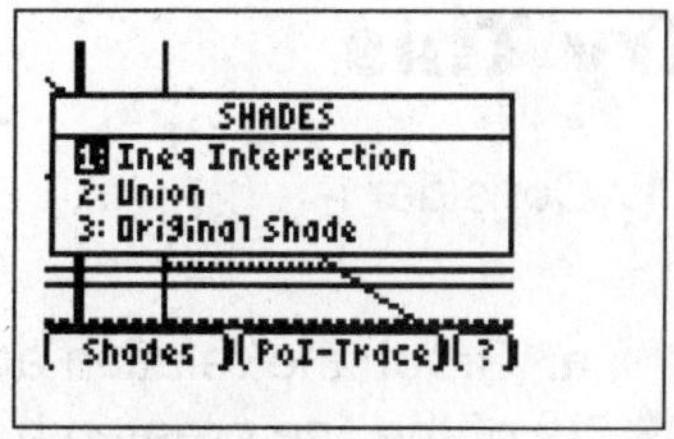

You need to set the Xmax and Ymax to higher values to show the entire graph. Press WINDOW.

Scroll down to **Xmax** and enter **300** Then scroll the down to **Ymax** and enter **300**. Press GRAPH.

View your graph

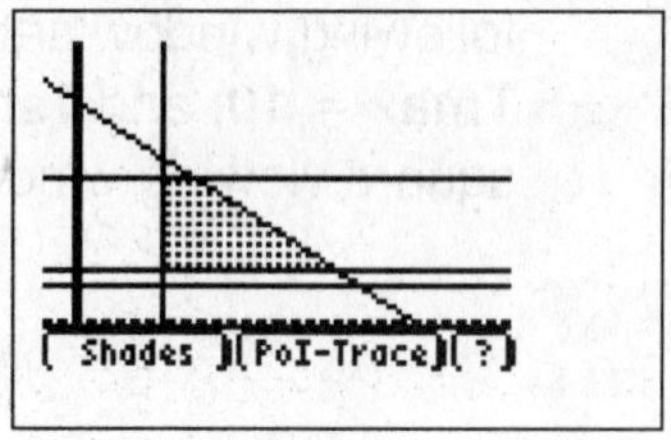

To shade only the feasible region, press ALPHA WINDOW and then **1**.

Now press ALPHA TRACE to find the four

intersection points **(80, 160), (60, 160), (200, 40), (60, 40)**.

By substituting the points into $P = 12.5x + 17.5y$, you can determine that the ordered pair (80, 160) produces the maximum revenue of $3800.00.

Try This

A local bird breeder can produce 80 chicks a year. He specializes in breeding doves and parakeets. The demand for his birds requires that he have between 15 and 40 doves and at least 20 parakeets. He sells the parakeets for $7.50 and sells the doves for 8.25.

1. What are the constraints and objective function?

2. How many of each bird should he produce?

3. What is the maximum revenue attainable?

Name ______________________________ Date ____________ Class ____________

LESSON **3-6**

Technology Lab Recording Sheet (p. 227)

Explore Parametric Equations

Try This

1. Consider $\begin{cases} x = 2t \\ y = \frac{1}{2}t \end{cases}$.

   a. Graph the relation and estimate the slope of the line created by the graph. Use the following window settings: **Tmin** = −10, **Tmax** = 10, and **Tstep** = 0.1, and use the square viewing window.

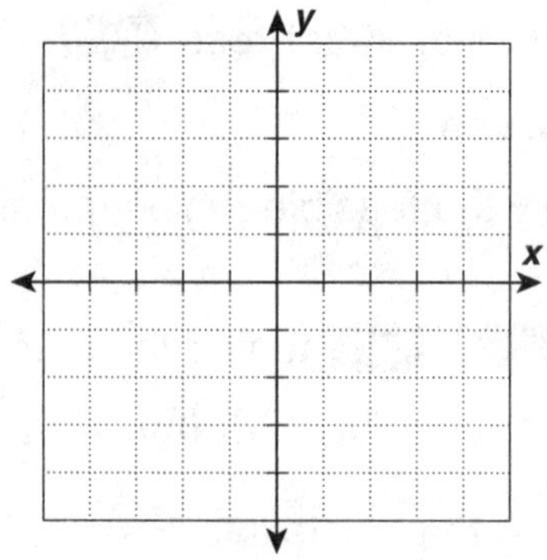

   b. Critical Thinking How does the slope of the line relate to the slope of the two parametric equations?

   ____

   ____

2. An airplane is traveling at a constant horizontal speed of 500 ft/s and is ascending at a constant rate of 50 ft/s. A model of the path of the airplane is given parametrically by $\begin{cases} x = 500t \\ y = 50t \end{cases}$ for t s.

   a. Graph the path of the airplane for $t = 20$ to $t = 180$ s.

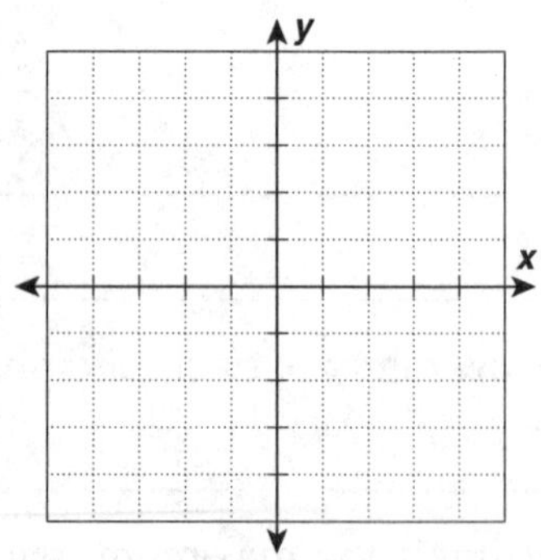

   Identify the location of the plane after 50 s.

   ____

   b. When will the plane reach an altitude of 5000 ft? (*Hint*: You may have to adjust your viewing window.)

   ____

3. **Critical Thinking** Create a graph of the line $y = x$ by using parametric equations.

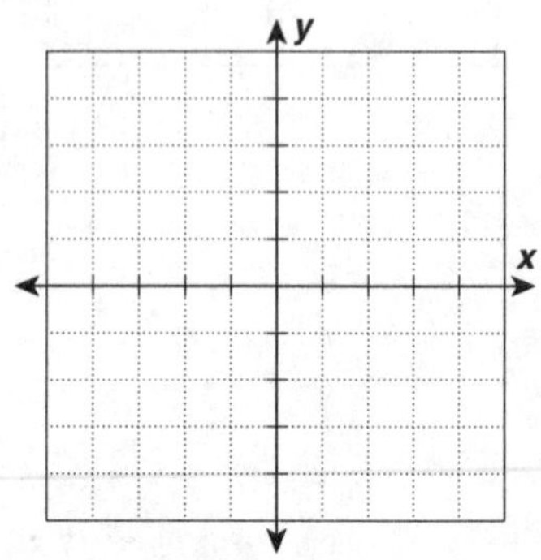

Name ______________________________ Date ____________ Class ____________

LESSON 4-5

Technology Lab Recording Sheet (p. 286)

Use Spreadsheets with Matrices to Solve Systems

Try This

1. Change the constants to −5 and 9, and solve the system by using a spreadsheet.

   The solution is: ____________

2. How can you check your answers by using the spreadsheet?

   ____

   ____

   ____

   ____

   ____

3. Solve a system you know to be inconsistent by using the spreadsheet.

   ____

   Solve a system you know to be dependent.

   ____

   How can you tell from the spreadsheet whether a system is inconsistent or dependent?

   ____

   ____

   ____

   ____

   ____

Name ______________________________ Date ____________ Class ____________

LESSON 4-6

Technology Lab

Solving Systems of Linear Equations in Three Variables

Use with Lesson 4-6

Activity

At the vegetable market, Jerry bought 1 melon, 3 red apples, and 5 oranges for a total of \$6.75. Mark bought 3 melons, 2 red apples, and 6 oranges for \$9.15. Doug bought 2 melons and 4 oranges for \$5.10. Find the cost of each type of fruit.

Let *x* equal the cost of one melon. Let *y* equal the cost of one red apple. Let *z* equal the number of oranges. The system of equations that represents this situation is:

$$\begin{cases} 1x + 3y + 5z = 6.75 \\ 3x + 2y + 6z = 9.15 \\ 2x \qquad\quad + 4z = 5.10 \end{cases}$$

The augmented matrix for this system of equations is:

$$\left[\begin{array}{ccc|c} 1 & 3 & 5 & 6.75 \\ 3 & 2 & 6 & 9.15 \\ 2 & 0 & 4 & 5.10 \end{array}\right]$$

Use a spreadsheet to represent this matrix.

	A	B	C	D	E
1		x	y	z	
2	Julie	1	3	5	6.75
3	Mark	3	2	6	9.15
4	Doug	2	0	4	5.1

Now solve by using row reduction on the spreadsheet.

The value in B2 is already a 1 so you do not need to convert cell B2. The value of cell B3 is 3 and should be converted to a zero. Multiply Row 1 by −3 and add to row 2. In cell B9, use the formula **=(B2*−3+B3)** to complete this row reduction. To convert the entire row, right click on the desired cell and drag it over the entire row. This will distribute the formula over the entire row.

7	Multiply R1 by –3 and add to R2				
8		1	3	5	6.75
9		0	–7	–9	–11.1
10		2	0	4	5.1
11					

The value of cell **B10** is 2 and must be converted to a zero. Multiply Row 1 by −2 and add to row 3. In cell **B15** use the formula **=(B8*−2+B10)** to complete this row reduction.

12	Multiply R1 by –2 and add to R3				
13		1	3	5	6.75
14		0	–7	–9	–11.1
15		0	–6	–6	8.4

The value of cell **C14** is −7 and must be converted to a 1. In cell **C19** use the formula **=(C14/−7)** to complete this row reduction.

LESSON **4-6**

Technology Lab

Solving Systems of Linear Equations in Three Variables continued

17	Multiply R2 by 1/–7				
18		1	3	5	6.75
19		0	1	1.28571	1.58571
20		0	–6	–6	–8.4

The value in cell **B20** is −6 and must be converted to a 0. In cell **B25** use the formula **=(C19*6+C20)** to complete this row reduction.

22	Multiply R2 by 6 and add to R#				
23		1	3	5	6.75
24		0	1	1.28571	1.58571
25		0	0	1.71429	1.11429

Cell **B25** is 1.71429 and must be converted to a 1. In cell **B30** use the formula **=(D25/1.71429)** to complete this row reduction.

27	Multiply R3 by 1/1.71429				
28		1	3	5	6.75
29		0	1	1.28571	1.58571
30		0	0	1	0.65

So, $z = 0.65$.

Now substitute $z = 0.65$ in for z in the equation to solve for x.

$$2x + 4(.65) = 5.10$$
$$x = 1.25$$

Now substitute $z = 0.65$ and $x = 1.25$ into one of the other equations and solve for y.

$$3(1.25) + 2y + 6(0.65) = 9.15$$
$$y = 0.75$$

Melons cost $1.25; red apples cost $0.75; oranges cost $0.65.

Try This

Use an Excel spreadsheet to solve the following problem.

The local movie theater has a matinee that offers a discount of 25% off of regularly priced tickets. The Johnson's buy 2 child tickets, 2 adult tickets, and 1 senior-citizen ticket for $33.75 at matinee prices. The Jackson family buys 1 adult and 1 senior-citizen ticket for $14.25 at matinee prices. The Myers family buys 3 child tickets and 1 adult and 2 senior-citizen tickets for $39.00 at matinee prices.

1. What is the cost of each type of ticket at matinee prices? ______________

2. What is the regular price of a ticket for a child? ______________

3. **Critical Thinking** How much money can a senior-citizen save if he or she bought 50 adult matinee priced tickets instead of the regular priced senior-citizen tickets?

____

Name ______________________ Date ____________ Class ____________

LESSON **5-1**

Technology Lab Recording Sheet (p. 314)

Explore Parameter Changes

Try This

Use your graphing calculator to compare the graph of each function to the graph of $f(x) = x^2$. Describe how the graphs differ.

1. $g(x) = (x - 4)^2$ ______________________

2. $g(x) = (x + 3)^2$ ______________________

3. $g(x) = -x^2$ ______________________

4. Make a Conjecture Use your graphing calculator to determine what happens when you change the value of h in the quadratic function $(x) = (x - h)^2$. Check both positive and negative values of h.

____

____

____

____

5. Make a Conjecture Use your graphing calculator to determine what happens when you change the value of a in the quadratic function $g(x) = ax^2$. Check values of a that are greater than 1 and values of a that are between 0 and 1.

____

____

____

____

Name ______________________ Date ____________ Class ____________

LESSON **5-3**

Technology Lab Recording Sheet (p. 332)

Explore Graphs and Factors

Try This

Graph each quadratic function and each of its linear factors. Then identify the *x*-intercepts and the axis of symmetry of each parabola.

1. $y = (x - 2)(x - 6)$ ____________

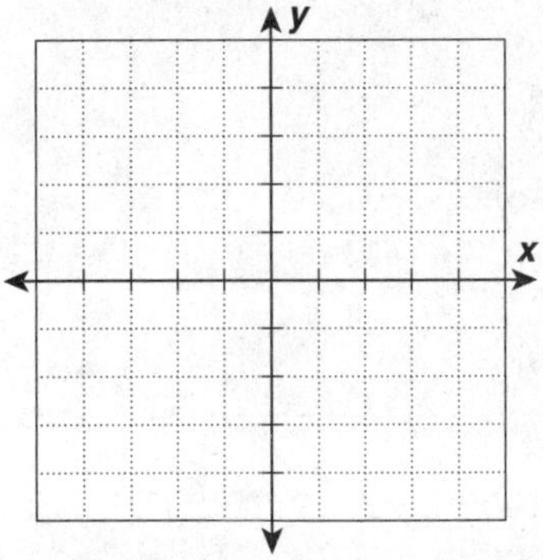

2. $y = (x + 3)(x - 1)$ ____________

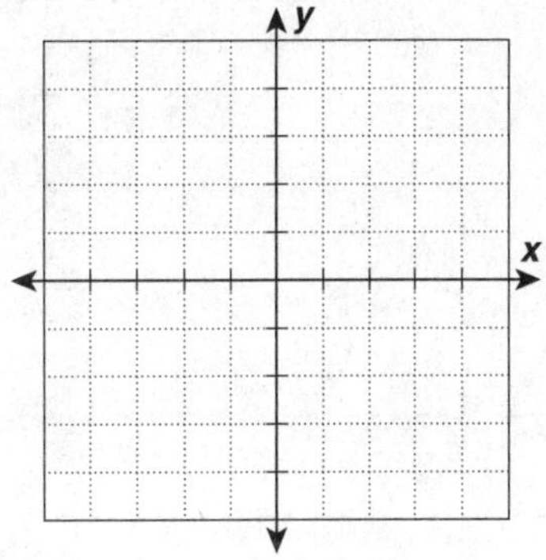

3. $y = (x - 5)(x + 2)$ ____________

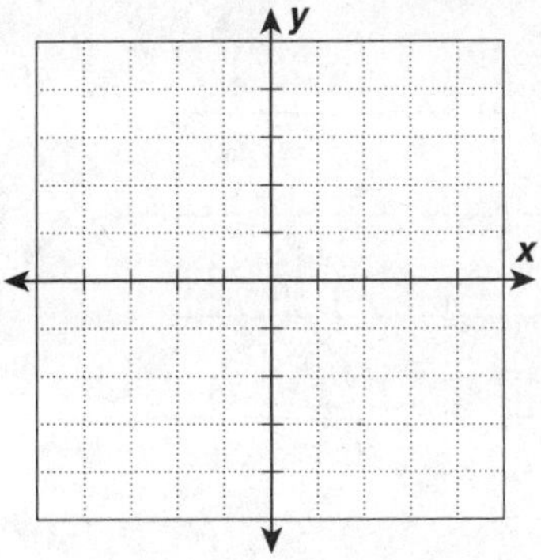

4. $y = (x + 4)(x - 4)$ ____________

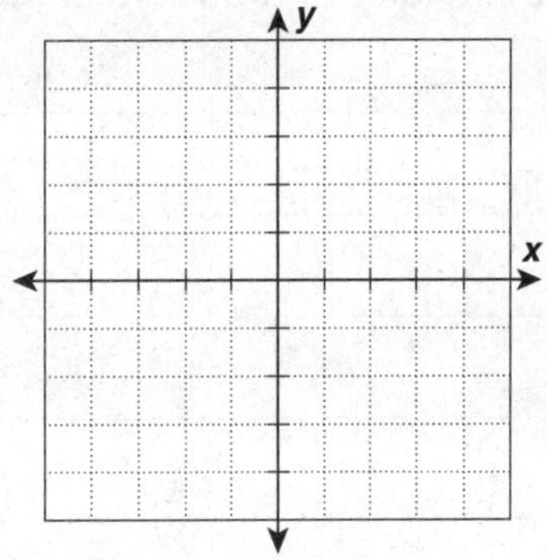

5. $y = (x - 5)(x - 5)$ ____________

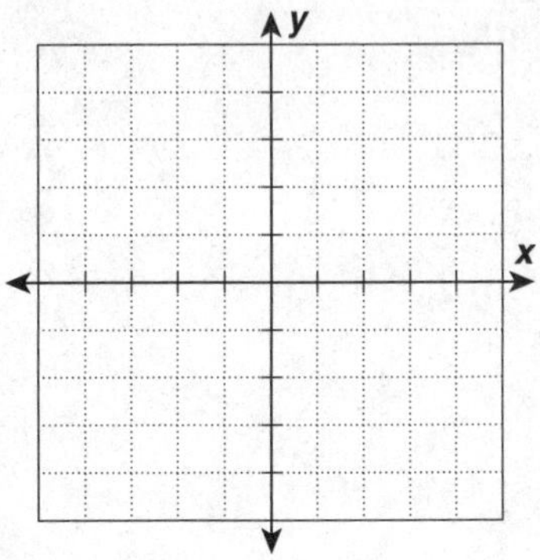

6. $y = (2x - 1)(2x + 3)$ ____________

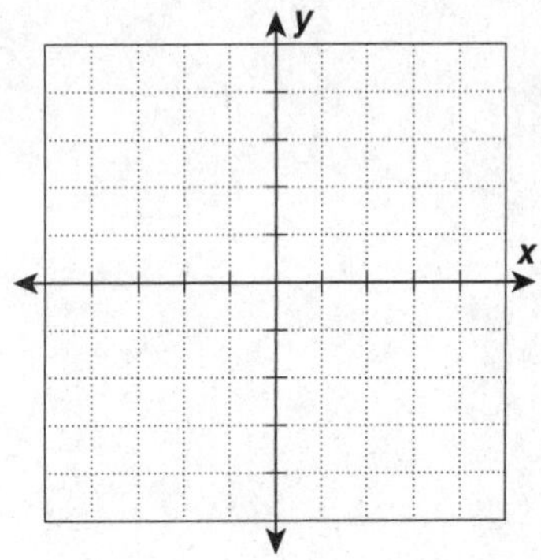

Name ______________________ Date __________ Class __________

LESSON 5-3

Technology Lab

Explore Graphs and Factors continued

7. **Critical Thinking** Use a graph to determine whether the quadratic function $y = 2x^2 + 5x - 12$ is the product of the linear factors $2x - 3$ and $x + 4$.

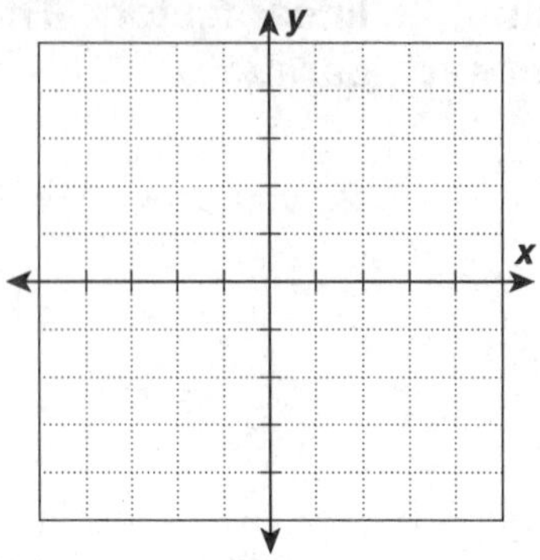

8. **Make a Conjecture** Make a conjecture about the linear factors, x-intercepts, and axis of symmetry of a quadratic function.

Name ______________________ Date __________ Class __________

LESSON 5-8

Technology Lab

Height and Time for a Bouncing Ball

Picture a bouncing ball. Between impacts with the floor, the ball rises and slows, then descends and speeds up. For any particular bounce, if the ball's height is plotted as a function of time, the resulting graph has a parabolic shape. In other words, the relationship between height and time for a single bounce of a ball is quadratic. This relationship is expressed mathematically as

$$y = ax^2 + bx + c$$

where y represents the ball's height at any given time x. Another form of a quadratic equation is

$$y = a(x - h)^2 + k$$

where h is the x-coordinate of the vertex, k is the y-coordinate of the vertex, and a is a parameter. This way of writing a quadratic is called the *vertex form.*

In this activity, you will record the motion of a bouncing ball using a Motion Detector. You will then analyze the collected data and model the variations in the ball's height as a function of time during one bounce using both the general and vertex forms of the quadratic equation.

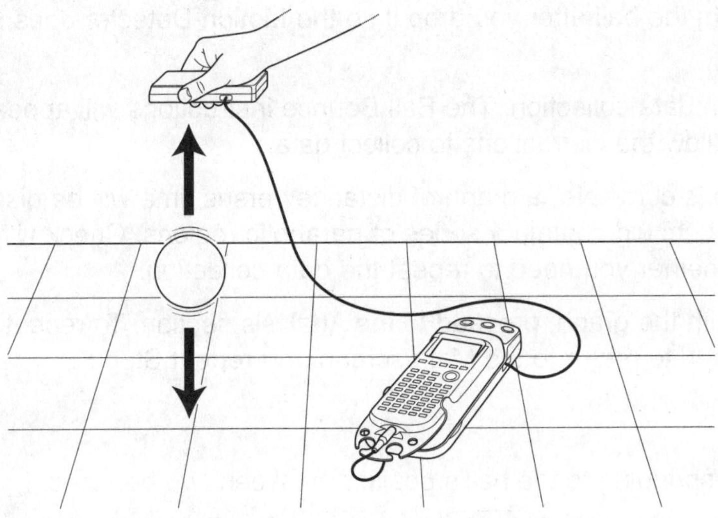

OBJECTIVES

- Record height versus time data for a bouncing ball.
- Model a single bounce using both the general and vertex forms of the parabola.

MATERIALS

TI-84 Plus graphing calculator
CBR 2 or Go! Motion and direct calculator cable
or Motion Detector and data-collection interface
EasyData application
ball (racquetball or basketball size)

Name ______________________________ Date __________ Class __________

LESSON **5-8**

Technology Lab

Height and Time for a Bouncing Ball continued

Activity

1. Set up the Motion Detector and calculator.
   a. Open the pivoting head of the Motion Detector. If your Motion Detector has a sensitivity switch, set it to Normal as shown.
   b. Turn on the calculator and connect it to the Motion Detector. (This may require the use of a data-collection interface.)
2. Position the Motion Detector about 1.5 m above the floor, so that the disc is pointing straight downward.
3. Set up EasyData for data collection.
   a. Start the EasyData application, if it is not already running.
   b. Select (File) from the Main screen, and then select **New** to reset the application.
   c. Select (Setup) from the Main screen, and then select **Ball Bounce**.
4. Practice dropping the ball so that it bounces straight up and down beneath the Motion Detector. Minimize the ball's sideways travel. Dropping the ball from about waist high works well. The ball must never get closer than 0.5 m from the detector. Be sure to pull your hands away from the ball after you drop it so the Motion Detector does not detect your hands.
5. Select (Start) to begin data collection. The Ball Bounce instructions will appear on the calculator screen. Follow the instructions to collect data.
6. When data collection is complete, a graph of distance versus time will be displayed. Examine the graph; it should contain a series of parabolic regions. Check with your teacher if you are not sure whether you need to repeat the data collection.

   If you are satisfied with the graph, proceed to the Analysis section. To repeat data collection, select (Main) to return to the Main screen and repeat Step 5.

ANALYSIS

1. Select the data corresponding to the ball's position between two bounces.
   a. From the distance graph, select (Anlyz), and then select **Select Region...** from the menu.
   b. If a warning is displayed on the screen; select (OK) to begin the region selection process.
   c. Use the ◀ and ▶ keys to move the cursor to the left edge of the parabolic region of your choice.
   d. Select (OK) to mark the left bound.
   e. Use the ◀ and ▶ keys to move the cursor to the right edge of the same parabolic region.
   f. Select (OK) to select the region. After a moment you will see the selected portion of your graph filling the width of the screen.

Name ______________________________ Date __________ Class __________

LESSON 5-8

Technology Lab

Height and Time for a Bouncing Ball continued

2. Trace across the graph to determine the *x*- and *y*-coordinates of the vertex of the parabola (in this case, the maximum point on the curve). Round these values to the nearest hundredth, and record them in the first Data Table on the Data Collection and Analysis sheet.

3. The model from physics for the distance versus time motion of a ball in free fall is quadratic. Fit a quadratic model $y = a(x - h)^2 + k$ to your data. In this vertex form, h is the *x*-coordinate of the vertex, k is the *y*-coordinate of the vertex, and a is the parameter.

   Since you have values for the parameters h and k of your model, you can try plotting the model using a guess for the a parameter. First, enter your model equation for graphing.

   a. Press [Y=].

   b. Press [CLEAR] to remove any existing equation.

   c. Enter the vertex form of the parabola using your values for the parameters h and k. For example, if your equation is $y = A(x - 4)^2 + 5$, enter $A*(x-4)^2 + 5$ in the Y1 field.

   d. Press [2nd] [QUIT] to return to the home screen.

4. Next, set a value for the parameter a, and then look at the resulting graph. To obtain a good fit, try several values for a. Use the steps below to store different values to the parameter a. Start with $a = 1$. Experiment until you find one that provides a good fit for the data. *Test both positive and negative values for a.*

   a. Enter a value for the parameter a. Press [STO▶] A [ENTER] to store the value in the variable A.

   b. Press [GRAPH] to see your data with the model graph superimposed.

   c. Press [2nd] [QUIT] to return to the home screen.

   Use your optimized value for a and the values of h and k you determined earlier to complete the vertex form of the equation.

5. It is also possible to express any quadratic function in the standard form of $y = ax^2 + bx + c$, where the coefficient a is the same as the coefficient you just found for the vertex form, and b and c are other parameters related to the h and k you already know. To determine the coefficients b and c, expand the vertex form of your equation and collect like terms.

   Round the corresponding values of a, b, and c to the nearest tenth and record them in the middle column of the second Data Table on the *Data Collection and Analysis* sheet.

6. Another way to determine the parameters is to use the calculator to perform a quadratic regression on the data to determine the parabola that fits the data best.

   a. Press [STAT] and use the cursor keys to highlight CALC.

   b. Press the number adjacent to QuadReg to copy the command to the home screen.

   c. Press [2nd] [L1] [,] [2nd] [L6] [,] to enter the lists containing the data.

   d. Press [VARS] and use the cursor keys to highlight Y-VARS.

   e. Select Function by pressing [ENTER].

   f. Press [ENTER] to copy Y1 to the expression.

Name ________________________ Date ____________ Class ____________

LESSON 5-8

Technology Lab

Height and Time for a Bouncing Ball continued

g. On the home screen, you will now see the entry QuadReg L1, L6, Y1. This command will perform a quadratic regression with L1 as the *x*-values and L6 as the *y*-values. The resulting regression curve will be stored in equation variable Y1. Press ENTER to perform the regression.

Copy the parameters *a*, *b* and *c* that appear on your calculator screen to the third column of the second Data Table.

h. Press GRAPH to see your graph.

DATA TABLE

Vertex	
***x*-coordinate**	***y*-coordinate**

Parameters	Values calculated from vertex form	Values from regression
a		
b		
c		

Try This

1. In this activity, the ball bounced straight up and down beneath the detector, yet the plot you see might seem to depict a ball that is moving sideways as it bounces up and down. Explain why the graph looks the way it does.

____

____

____

2. Write the vertex form of the parabola from Step 6.

____

3. Are the values of *a*, *b*, and *c* in the quadratic regression consistent with the values you determined earlier?

____

____

Name ______________________ Date __________ Class __________

LESSON 5-8

Technology Lab

Height and Time for a Bouncing Ball continued

4. Describe how the parameter a affects the graph of $y = a(x - h)^2 + k$. Specifically, how do the magnitude of a and the sign of a change the graph?

____

____

____

____

5. Suppose you had chosen the parabolic section for the bounce just to the right of the one you actually used in this activity. Describe how the parameters h and k would change, if at all, if this different parabolic section were to be fit with the equation $y = a(x - h)^2 + k$.

____

____

____

____

Name ______________________ Date ___________ Class ___________

LESSON 6-7

Technology Lab Recording Sheet (p. 452)

Explore End Behavior

Try This

1. Consider the functions $g(x) = 3x^3 - 2x^2 + x + 4$, $h(x) = \frac{1}{2}x^3 + 2x^2 + x - 9$ and $k(x) = 5x^3 - 8x^2 - 2x + 1$

   a. What do the functions $g(x)$, $h(x)$, and $k(x)$ have in common?

   ____

   ____

   b. Graph $g(x)$, $h(x)$, and $k(x)$ on your graphing calculator, and describe the end behavior of each.

   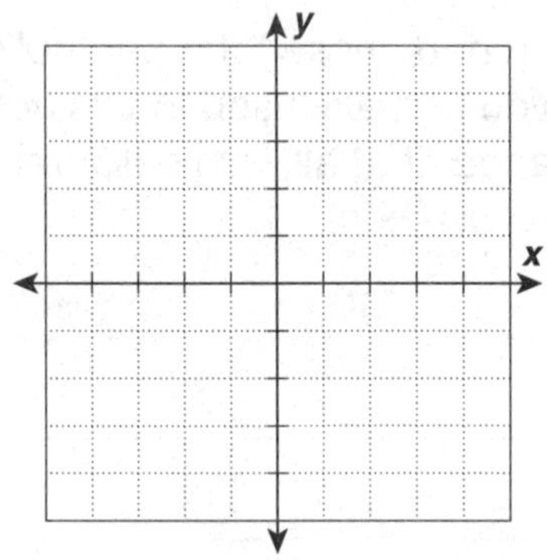

   ____

   ____

   ____

   ____

   c. Make a Conjecture What can you say about the end behavior of functions of the same type as $g(x)$, $h(x)$, and $k(x)$?

   ____

   ____

2. Consider the functions $a(x) = -3x^3 - 2x^2 + x + 4$, $b(x) = -\frac{1}{2}x^3 + 3x^2 + x - 9$, and $c(x) = -5x^3 - 8x^2 - 2x + 1$.

   a. What do the functions $a(x)$, $b(x)$, and $c(x)$ have in common?

   ____

   ____

   b. Graph $a(x)$, $b(x)$, and $c(x)$ on your graphing calculator, and describe the end behavior of each.

   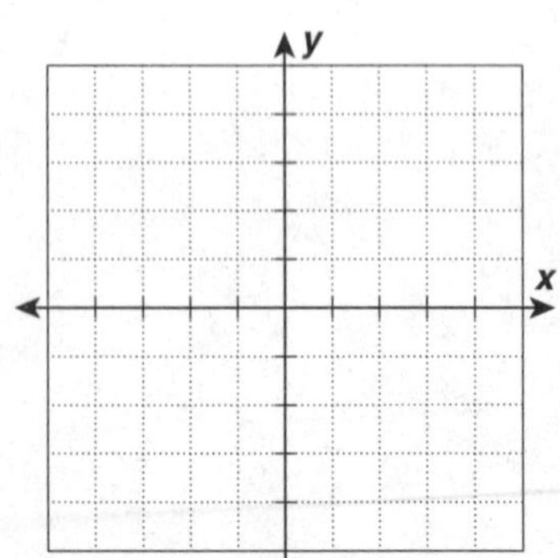

   ____

   ____

   ____

   ____

Name ______________________ Date __________ Class __________

LESSON 6-7

Technology Lab Recording Sheet

Explore End Behavior continued

c. Make a Conjecture What can you say about the end behavior of functions of the same type as $a(x)$, $b(x)$, and $c(x)$?

____

____

3. Consider the functions $p(x) = 3x^4 - x^2 + x + 4$, $r(x) = \frac{1}{2}x^4 + 3x^3 + x - 9$, and $s(x) = 5x^4 - 8x^3 - 2x^2 + 1$.

a. What do the functions $p(x)$, $r(x)$, and $s(x)$ have in common?

____

____

b. Graph $p(x)$, $r(x)$, and $s(x)$ on your graphing calculator, and describe the end behavior of each.

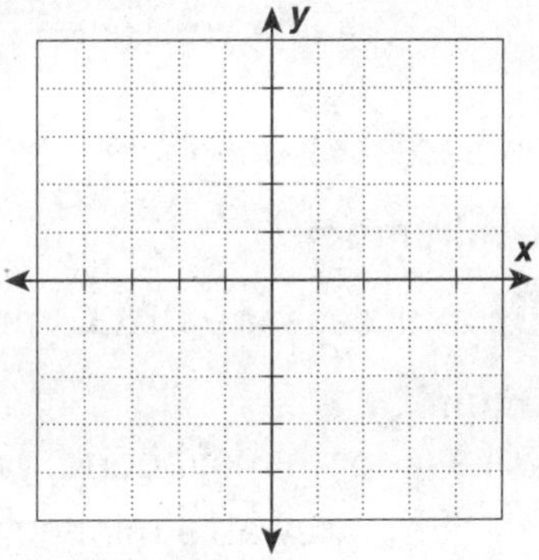

____

____

____

____

c. Make a Conjecture What can you say about the end behavior of functions of the same type as $p(x)$, $r(x)$, and $s(x)$?

____

____

4. Critical Thinking Compare your conjectures from Problems 1c, 2c, and 3c. What are the characteristics of a function that seem to affect the function's end behavior?

____

____

Name ______________________ Date ____________ Class ____________

LESSON **6-7**

Technology Lab
Graphing Polynomial Functions

Use with Lesson 6-7

You can use a graphing calculator to estimate the local maxima and minima of polynomial functions.

Activity

Graph the function $g(x) = 3x^3 - 15x + 7$ on a calculator, and estimate the local maxima and minima.

Begin by pressing **Y**. For Y_1 input $\mathbf{3x^3-15x+7}$. Now press **GRAPH**.

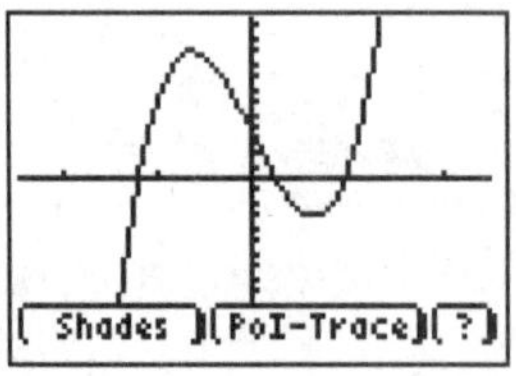

Step 1 Graph.

The graph appears to have one local maximum and one local minimum.

Step 2 Find the minimum.

Press **2nd** **TRACE** to access the **CALC** menu.

Choose 3: **minimum**

The calculator will ask for the leftbound; press the arrow key until the curser is approximately left of the viewable minimum and then press **ENTER**.

The calculator will ask for the rightbound; press the arrow key until the curser is approximately left of the viewable minimum and then press **ENTER**.

The calculator will ask whether to guess.

Press **ENTER**.

The local minimum is **5.80929**

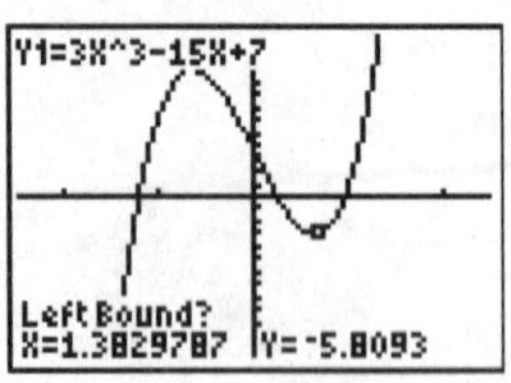

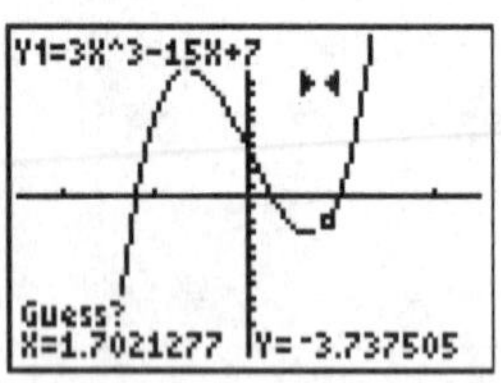

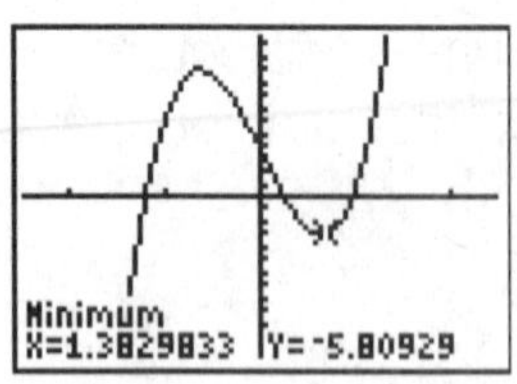

Name ______________________________ Date __________ Class ___________

LESSON 6-7

Technology Lab

Graphing Polynomial Functions continued

Step 3 Find the Maximum.

Press 2nd TRACE to access the **CALC menu.**
Choose 3: **maximum.**
The calculator will ask for the leftbound; press the arrow key until the curser is approximately left of the viewable maximum and then press ENTER.
The calculator will ask for the rightbound; press the arrow key until the curser is approximately right of the viewable maximum and then press ENTER.
The calculator will ask whether to guess.
Press ENTER.
The local maximum is **19.745**

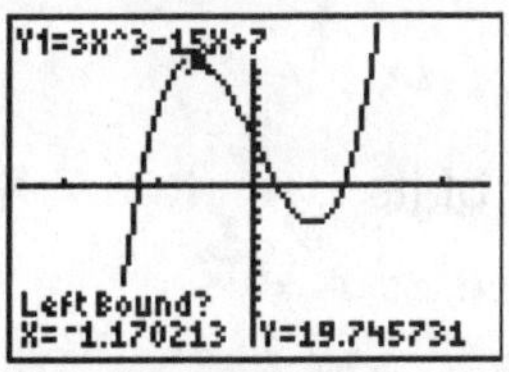

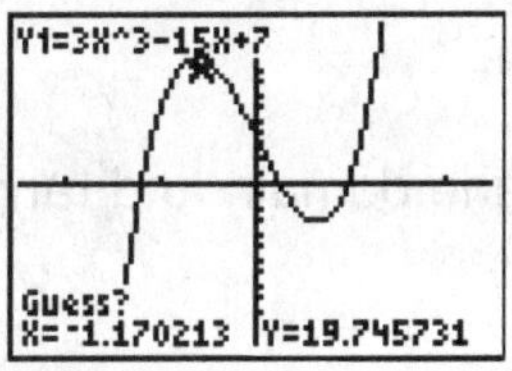

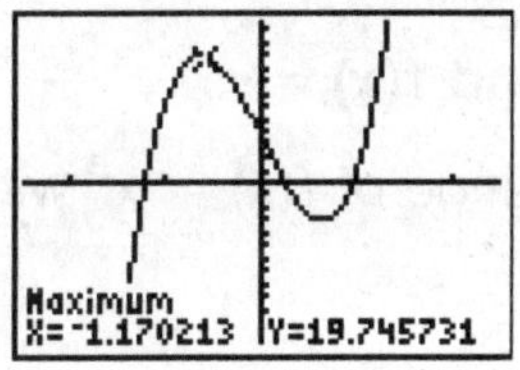

Try This

Graph each function on a calculator, and estimate the local maximum and minimum of the following functions.

1. $d(x) = 3x^3 - 2x + 12$

Maximum ______________________________

Minimum ______________________________

2. $d(x) = x^3 - 2x + 5$

Maximum ______________________________

Minimum ______________________________

Name ______________________ Date __________ Class __________

LESSON **7-2**

Technology Lab Recording Sheet (p. 497)

Explore Inverses of Functions

Try This

Graph $f(x) = x^2$, its inverse, and $f(x) = x$.

1. Compare the domain and range of $f(x) = x^2$ with the domain and range of its inverse.

____

Is the inverse of $f(x) = x^2$ a function? Explain why or why not.

____

Graph $f(x) = x^3$, its inverse, and $f(x) = x$.

2. Compare the domain and range of $f(x) = x^3$ with the domain and range of its inverse.

____

Is the inverse of $f(x) = x^3$ a function? Explain why or why not.

____

3. **Make a Conjecture** Make a conjecture about the relationship between the domain and range of a function and its inverse.

____

____

4. **Make a Conjecture** Make a conjecture about the relationship of a function and its inverse to the line $f(x) = x$.

____

____

Name ______________________________ Date ____________ Class ____________

LESSON **7-6**

Technology Lab Recording Sheet (p. 530)

Explore the Rule of 72

Try This

1. Complete cells A3 through A20 with interest rates from 3.5% to 12%.

2. Copy and paste the formula from cell B2 into cells B3 through B20.

3. Copy and paste the formula from cell C2 into cells C3 through C20.

	A	B	C
1			
2			
3			
4			
5			
6			
7			
8			
9			
10			
11			
12			
13			
14			
15			
16			
17			
18			
19			
20			

4. What is the lowest interest rate that gives you a doubling time less than 9 years?

____

5. **Make a Conjecture** The rule of 72 is a rule used by investors to estimate the time it will take for an investment to double at a certain rate. Explain why.

____

____

6. **Critical Thinking** Write an equation for the rule of 72. ____________

Name ______________________ Date __________ Class __________

LESSON 7-8

Technology Lab

How Hot Objects Cool

When you have a hot drink, you know that it gradually cools off. Newton's law of cooling provides us with a model for cooling. It states that the temperature difference T_{diff} between a hot object and its surroundings decreases exponentially with time.

$$T_{diff} = T_0 e^{-kt}$$

In the model T_0 is the initial temperature difference, and k is a positive constant.

In this activity you will use an EasyTemp temperature probe to collect data as the warmed probe cools. You can then fit several mathematical models to the data.

OBJECTIVES

- Record temperature versus time cooling data.
- Model cooling data with an exponential function.

MATERIALS

TI-84 Plus or TI-84 Plus Silver Edition graphing calculator
Vernier EasyTemp
Vernier EasyData application
hot water

Activity

1. Turn on your TI-84 Plus (or TI-84 Plus Silver Edition) graphing calculator and make sure that it is on the home screen. Plug the EasyTemp probe into the USB port of the graphing calculator. The EasyData application will automatically start and the Main screen will be displayed.
2. Select (File) from the Main screen, and then select **New**.
3. Obtain a cup of hot water at 45 to 55 °C. Place the EasyTemp probe in the water, and wait about 20 seconds for the probe to reach the temperature of the water.
4. Remove the probe from the water and rest it on the edge of a table. Do not let anything touch the tip of the probe. Select (Start) from the Main screen. Data collection will run for three minutes, after which a graph of temperature versus time will be displayed.
5. Newton's law of cooling models the temperature difference between the warm object and its surroundings. As an estimate of the room temperature, use the minimum temperature on the graph. Inspect the graph to determine its minimum temperature reading.
   a. Trace along the graph with the ▶ key and determine the minimum temperature reached. Round this temperature down to the next whole degree (as in 24.54 → 24) and record the value in the Data Table.
   b. Select (Main) to return to the Main screen.
   c. Select (Quit) from the Main screen. Select (OK) to exit the EasyData application.

Name ________________________________ Date __________ Class __________

LESSON 7-8

Technology Lab

How Hot Objects Cool continued

ANALYSIS

1. Since the model for Newton's law of cooling uses the difference between the temperature of the warm object and its surroundings, subtract the minimum temperature from the measured temperature before comparing data to the model.
   a. Press [2nd] [L2].
   b. Press [−] then key in the (minimum temp), where (minimum temp) is the numerical value you entered in the Data Table.
   c. Press [STO▸], and press [2nd] [L2] a second time. Your expression will look like L2 − 24 → L2, depending on your particular minimum temperature. Press [ENTER] to perform the calculation.
2. Display your graph of temperature difference versus time.
   a. Press [ZOOM].
   b. Press [▼] until ZoomStat is highlighted; press [ENTER] to display a graph with the x and y ranges set to fill the screen with data.
   c. Press [TRACE] to determine the coordinates of a point on the graph using the cursor keys.
3. Plot an exponential model of $y = Te^{-kx}$ to the data. Use these new variable names to match what the calculator can use. You can determine the value for T (the initial temperature difference T_0) by tracing the graph to the point at $x = 0$. Record this parameter as T in the Data Table.
4. Because you have values for the parameter T of the model, you can try plotting the model using a guess for the k parameter. First, enter the model equation for graphing.
   a. Press [Y=].
   b. Press [CLEAR] to remove any existing equation.
   c. Enter the exponential model using the value for the parameter T. For example, if the equation is $y = 25e^{-kx}$, enter 25*e^(−K*x) on the Y1 line.
   d. Press [◄] until the icon to the left of Y1 is blinking. Press [ENTER] until a bold diagonal line is shown which will display your model with a thick line.
   e. Press [2nd] [QUIT] to return to the home screen.
5. Set a value for the parameter k, and then look at the resulting graph. To obtain a good fit, you will need to adjust the value of k. Use the steps below to store different values to the parameter k, starting with $k = 0.01$.
   a. Enter a value for the parameter k. Press [STO▸] K [ENTER] to store the value in the variable K.
   b. Press [GRAPH] to see the data with the model graph superimposed. The model line is bold.
   c. Press [2nd] [QUIT] to return to the home screen.

   Experiment until you find one that provides a good fit for the data. Record the k value that works best in the Data Table.

Name ________________________________ Date ___________ Class ___________

LESSON 7-8

Technology Lab

How Hot Objects Cool continued

6. You can also use the calculator to fit an exponential to the data. The calculator uses the exponential form of $y = ab^x$. This is slightly different from the form you used in the modeling of the previous step, but it is still an exponential.

   a. Press STAT and use the cursor keys to highlight CALC.

   b. Press the number adjacent to ExpReg to copy the command to the home screen.

   c. Press 2nd [L1] , 2nd [L2] , to enter the lists containing the data.

   d. Press VARS and use the cursor keys to highlight Y-VARS.

   e. Select Function by pressing ENTER.

   f. Press ENTER to copy Y1 to the expression.

   On the home screen, you will now see the entry ExpReg L1, L2, Y1. This command will perform an exponential regression with L1 as the *x* values and L2 as the *y* values. The resulting regression equation will be stored in equation variable Y1. Press ENTER to perform the regression. Copy the parameters a and b that appear on the calculator screen to the Data Table.

   g. Press GRAPH to see your graph.

DATA TABLE

Minimum Temperature (°C)		
$y = Te^{-kx}$	T	
	k	
$y = ab^x$	a	
	b	

Try This

1. Is the graph consistent with the model of a decreasing exponential? In what way?

____

____

2. Record the model equation from Step 5 of the Analysis section here.

____

3. How does the regression fit compare to the model? How can both models fit the data well when the fit parameters are different?

____

____

Name ______________________________ Date ____________ Class ____________

LESSON 7-8

Technology Lab

How Hot Objects Cool continued

4. Why are the values for a and T similar, while b and k are quite different? Use the model equations to find the relationship between k and b so you can compare the value to that of the calculator.

____

____

5. When $t = 0$, what is the value of e^{-kt}?

____

6. When t is very large, what is the value of the temperature difference? What is the temperature of the sensor at this time?

____

____

7. What could you do to the experimental apparatus to decrease the value of k in another run? What quantity does k measure?

____

____

8. Use either the model equation or the regression equation to predict the time it takes the sensor to reach a temperature 1°C above room temperature.

____

____

9. If the starting temperature difference is cut in half, does it take half as long to get to 1°C above room temperature? Explain.

____

____

Name ______________________________ Date __________ Class __________

LESSON 8-1

Technology Lab

Light at a Distance: Distance and Light Intensity

While traveling in a car at night, you may have observed the headlights of an oncoming vehicle. The light starts as a dim glow in the distance, but as the vehicle gets closer, the brightness of the headlights increases rapidly. This is because the light spreads out as it moves away from the source. As a result, light intensity decreases as the distance from a typical light source increases. What is the relationship between distance and intensity for a simple light bulb?

In this activity you can explore the relationship between distance and intensity for a light bulb. You will record the intensity at various distances between a Light Sensor and the bulb. The data can then be analyzed and modeled mathematically.

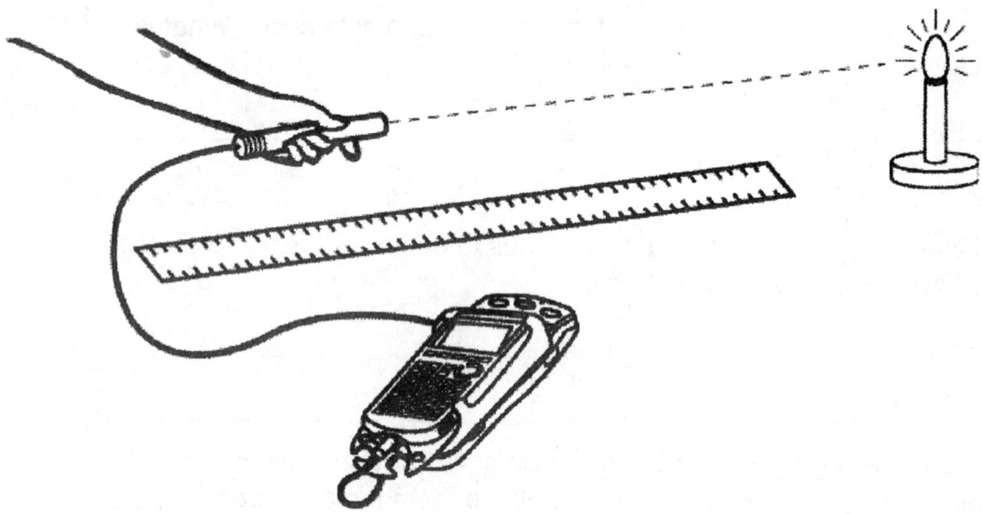

OBJECTIVES

- Collect light intensity versus distance data for a point light source.
- Compare data to an inverse-square model.
- Compare data to a power law model.
- Discuss the difference between an inverse-square model and a power law model.

MATERIALS

TI-84 Plus graphing calculator
EasyData application
data-collection interface
Light Sensor
meter stick or tape measure
dc-powered point light source

LESSON 8-1

Technology Lab

Light at a Distance: Distance and Light Intensity continued

Activity

1. Arrange the equipment. There must be no obstructions between the bulb and the Light Sensor during data collection. Remove any surfaces near the bulb, such as books, people, walls or tables. There should be no reflective surfaces behind, beside, or below the bulb. The filament and Light Sensor should be at the same vertical height. This makes the bulb look more like a point source of light as seen by the Light Sensor. While you are taking intensity readings, the Light Sensor must be pointed directly at the light bulb.
2. Set up the Light Sensor for data collection.
   a. Turn on the calculator.
   b. If you are using the Vernier Light Sensor, set it to 0–600 lux for a small light source, or 0–6000 lux for a larger light source.
   c. Connect the Light Sensor, data-collection interface, and calculator.
3. Set up EasyData for data application.
   a. Start the EasyData application, if it is not already running.
   b. Select (File) from the Main screen, and then select **New** to reset the application.
   c. Select (Setup) from the Main screen, and then select **Events with Entry**.
4. Dim the lights to darken the room. A very dark room is critical to obtain good results.
5. Hold the Light Sensor about 10 cm from the light bulb filament. Move the sensor away from the bulb and watch the displayed intensity values on the calculator screen.
6. To account for the particular brightness of your light source, choose a starting distance that gives a reading less than the maximum reading for your sensor (600 or 6000 lux for the Vernier sensor, or 1 for the TI sensor), but as large as possible. However, do not get any closer than 5 cm for small (<5 mm) bulbs, or 10 cm otherwise. Choose the starting distance, and enter it as X_L in the Data Table.
7. Again place the Light Sensor at your planned starting distance from the light bulb filament. **Important:** The distance must be measured carefully. Be sure you measure from the filament of the lamp to the sensor tip on the Light Sensor.
8. Select (Start) from the Main screen to prepare for data collection.
9. Wait for the value displayed on the calculator to stabilize. Select (Keep), and then enter the distance between the Light Sensor and the light source in meters on the calculator. Select (OK) to store the data pair.
10. Move the Light Sensor 1 cm farther away from the light source and repeat Step 9.
11. Continue moving the sensor in 1-cm increments until the readings fall to less than 10% of the initial reading, collecting data as before. After the final data point, select (Stop) to end data collection.
12. Inspect the graph of light intensity versus distance. Trace to read the *x* and *y* values of the left-most point, round the values to three significant figures, and record them as X_L and Y_L in the Data Table.

Name ______________________ Date __________ Class __________

LESSON 8-1

Technology Lab

Light at a Distance: Distance and Light Intensity continued

13. Select (Main) to return to the Main screen. Exit EasyData by selecting (Quit) from the Main screen and then selecting (OK).

ANALYSIS

1. Redisplay your graph outside of EasyData.
   a. Press [2nd] [STAT PLOT].
   b. Press [ENTER] to select Plot1 and then press [ENTER] again to select On.
   c. Press [ZOOM].
   d. Press [▼] until ZoomStat is highlighted; press [ENTER] to display a graph with the *x* and *y* ranges set to fill the screen with data.
2. Inspect your graph of the light intensity versus distance.
3. One model for light intensity holds that the intensity is proportional to the inverse square of the distance from a point light source; that is, a graph would be of the form $y = C/x^2$, where *C* is an adjustable parameter. Does your data follow this model? You can check it out by finding an approximate value for *C*, and then graphing the model with the data. First, enter the model equation.
   a. Press [Y=].
   b. Press [CLEAR] to remove any existing equation.
   c. Enter C/x^2 in the Y1 field.
   d. Press [2nd] [QUIT] to return to the home screen.
4. Set a value for the parameter *C* and then look at the resulting graph. To obtain a good fit, you will need to try several values for *C*. Use the steps below to determine an initial guess for the parameter. One way to find an approximate value for the parameter is to use the left-most point to calculate an initial value for *C*. Record this value in the Data Table.
   a. Enter a value for the parameter *C*. Press [STO▶] C [ENTER] to store the value in the variable C.
   b. Press [GRAPH] to see the data with the model graph superimposed.
   c. Press [2nd] [QUIT] to return to the home screen.
5. If the model is systematically high or low, you may want to adjust the value of C to improve the fit. As you did before, store a new value in C, and then display the graph and model. Once you have optimized the model, record the complete equation in the Data Table.
6. Another model can be used to compare to your data. The general power law of $y = ax^b$ may provide a better fit than the inverse square function, especially if the light source is not small or if there are reflections from walls or other surfaces. The difference between this new model and the inverse square model is that the exponent is not fixed at −2. Instead,

Name ________________________________ Date ____________ Class ____________

LESSON **8-1**

Technology Lab

Light at a Distance: Distance and Light Intensity continued

the exponent is now an adjustable parameter. The calculator can be used to automatically determine the parameters *a* and *b* in the general power law relation to the data.

a. Press STAT and use the cursor keys to highlight CALC.

b. Press ▼ repeatedly to scroll down to PwrReg. When it is highlighted, press ENTER to copy the command to the home screen.

c. Press 2nd [L1] , 2nd [L2] , to enter the lists containing the data.

d. Press VARS and use the cursor keys to highlight Y-VARS.

e. Select Function by pressing ENTER.

f. Press ENTER to copy Y1 to the home screen.

g. On the home screen, you will now see the entry PwrReg L1, L2, Y1. This command will perform a power law regression with L1 as the *x* and L2 as the *y* values. The resulting regression curve will be stored in equation variable Y1. Press ENTER to perform the regression. Use the parameters a and b, rounded to two significant figures, to write the power law model equation in the Data Table.

h. Press GRAPH to see the graph of your data and the power regression function.

DATA TABLE

Left-most point *X* value X_L	
Left-most point *Y* value Y_L	
Initial model parameter *C*	
Optimized inverse-square model $y = C/x^2$	
Power law model $y = ax^b$	

Try This

1. What is your prediction for the relationship between intensity and the distance to a light source?

____

____

____

____

2. Is the graph of the light intensity versus distance consistent with your earlier prediction?

____

____

____

____

Name ______________________ Date ____________ Class ____________

LESSON 8-1

Technology Lab

Light at a Distance: Distance and Light Intensity continued

3. How well does the inverse-square model $y = \frac{C}{x^2}$ fit experimental data?

____

____

4. How well does the power law model fit the data? Could it fit any better than the inverse-square model? Could it fit more poorly?

____

____

____

____

____

____

____

5. How would using a brighter light bulb affect the parameters *a*, *b*, and *C* in the two models?

____

____

____

____

Name ______________________ Date ___________ Class ___________

LESSON 8-4

Technology Lab Recording Sheet (p. 591)

Explore Breaks in Graphs

Try This

Use a graph and a table to identify the hole in the graph of each function.

1. $f(x) = \dfrac{(x - 2)(x + 3)}{x + 3}$

____

2. $g(x) = \dfrac{(x + 1)(x + 3)}{x + 1}$

____

3. $h(x) = \dfrac{x(x + 2)}{x + 2}$

____

4. Make a Conjecture Make a conjecture about where the holes in the graph of a rational function appear, based on the factors of the numerator and the denominator.

____

____

Use a graph and a table to identify the hole(s) in the graph of each function. Confirm your answer by factoring.

5. $f(x) = \dfrac{x^2 - 4x + 3}{x - 3}$

____

6. $g(x) = \dfrac{x^2 + x - 2}{x - 1}$

____

7. $h(x) = \dfrac{x^3 - x}{x^2 - 1}$

____

Name ______________________ Date ____________ Class ____________

LESSON **8-7**

Technology Lab

Graphing Radical Inequalities

Use with Lesson 8-7

Activity 1

Graph the inequality $y < \sqrt{x + 9}$.

Begin by pressing APPS and choose **Inequals** and press ENTER.

Press Y= and change the = sign to a < by pressing ALPHA and then WINDOW.

Now press 2nd and then x^2. Press X,T,θ,n + 9 and then press) before you press graph.

You must change the viewing window to see the entire graph.

Press WINDOW Change **Xmin=−15, Xmax=10 Xscl=1, Ymin=−5, Ymax=25, Yscl=5**. Now press GRAPH.

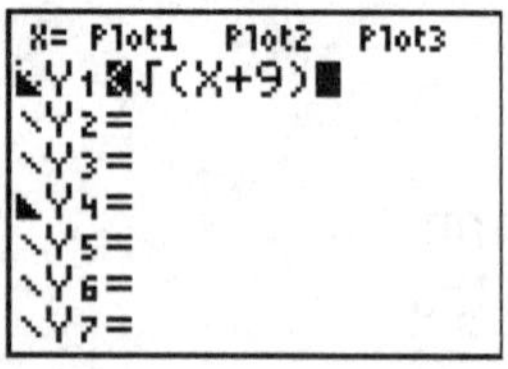

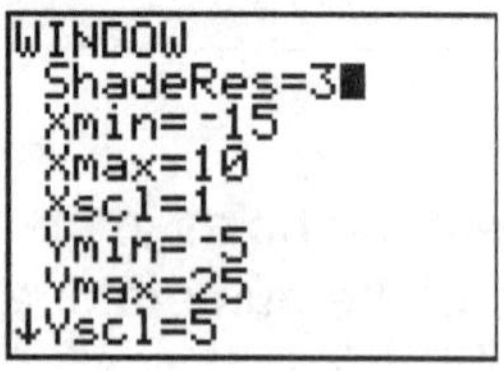

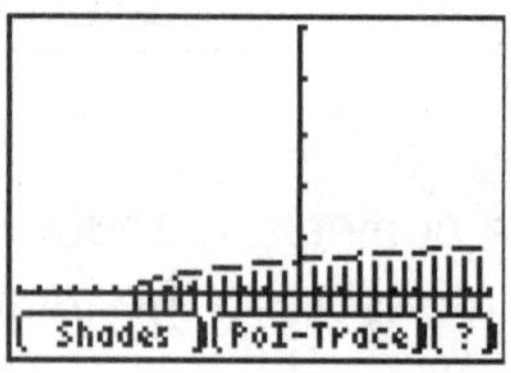

Try This

Graph each inequality.

1. $y < \sqrt{x} + 6$

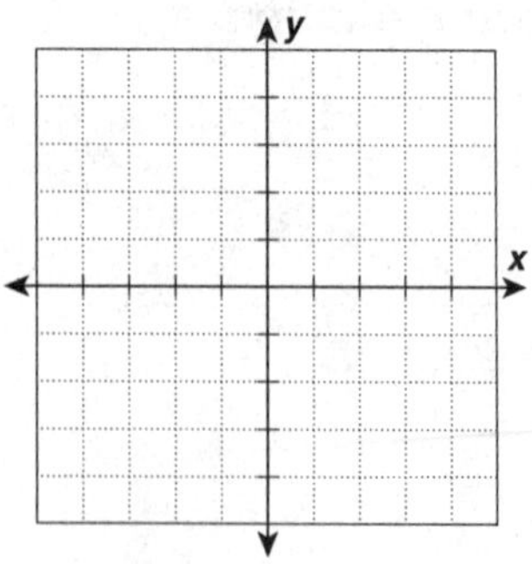

2. $y < \sqrt{x - 12}$

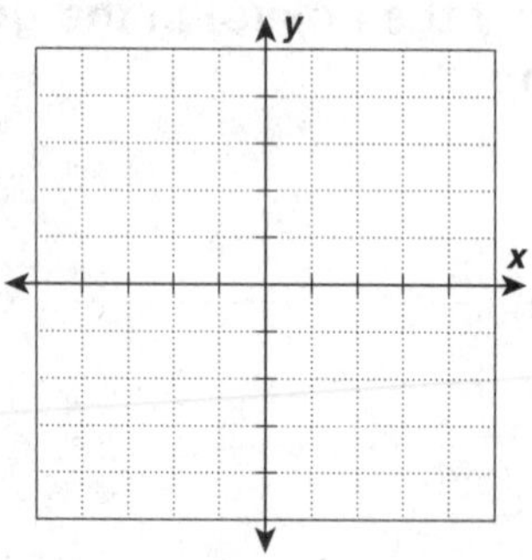

3. $y \geq \sqrt{x - 9}$

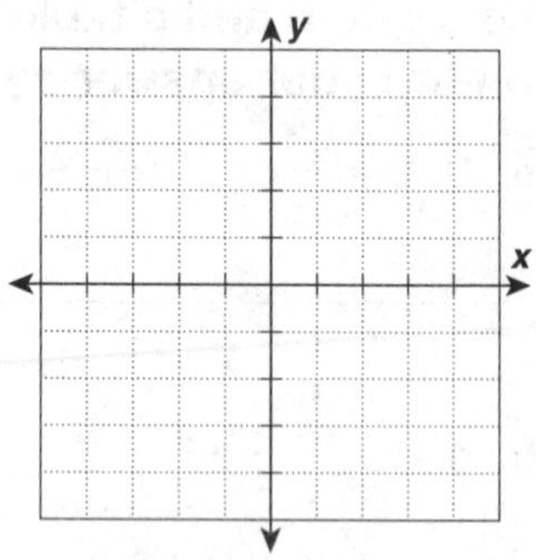

4. $y \geq \sqrt{x} - 8$

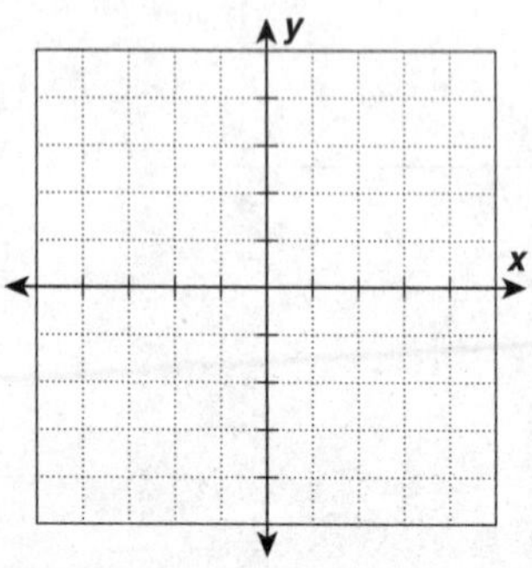

5. $y \leq \sqrt{x} - 2$

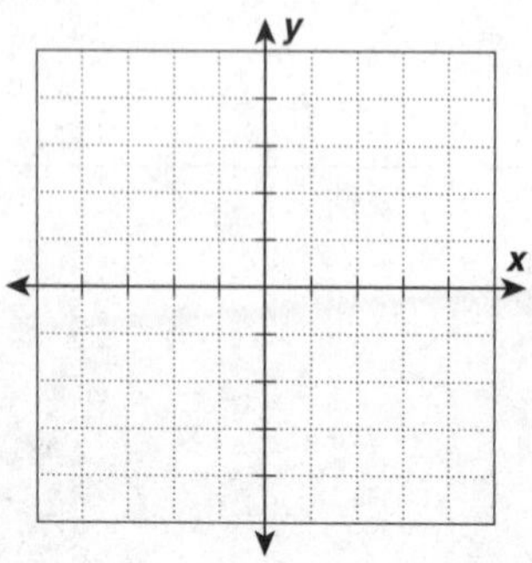

6. $y \leq \sqrt[3]{x} - 6$

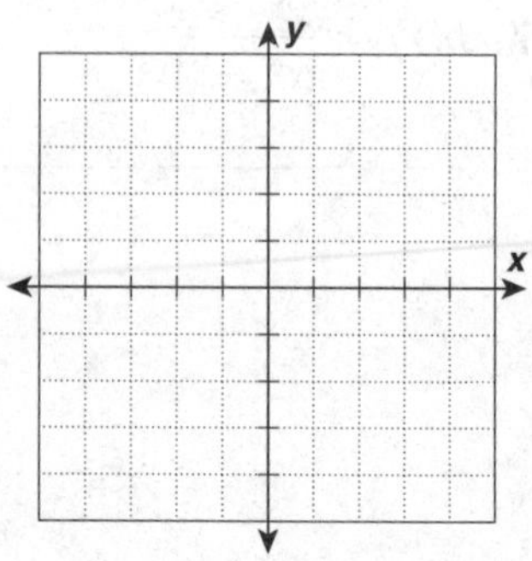

Name ______________________ Date __________ Class __________

LESSON 9-2

Technology Lab Recording Sheet (pp. 670–671)

Graph Piecewise Functions

Try This

Activity 1

Use logical tests to determine whether each statement is true or false.

1. $4 - 3 = 3 - 4$ ______________

2. $(-6)^4 \geq 1000$ ______________

3. $\frac{3}{16} < 0.1875$ ______________

4. $\frac{3}{16} > 0.1875$ ______________

5. Draw a Conclusion What conclusion can you make about $\frac{3}{16}$ and 0.1875 based on the answers to Problems 3 and 4?

____

Activity 2

6. $-2^2 < -2 < (-2)^2$ ______________

7. $(-2)^1 < (-2)^2 < (-2)^3$ ______________

8. $\sqrt{2} < 2 < 2^2$ ______________

9. $\sqrt{\frac{1}{2}} < \frac{1}{2} < \left(\frac{1}{2}\right)^2$ ______________

10. Make a Conjecture What conjecture can you make about the relationship between a number, its square, and its square root if the number is greater than 1?

____

What if the number is between 0 and 1?

____

Activity 3

Graph each piecewise function.

11. $g(x)\begin{cases} x & \text{if } x < 0 \\ -x & \text{if } x \geq 0 \end{cases}$

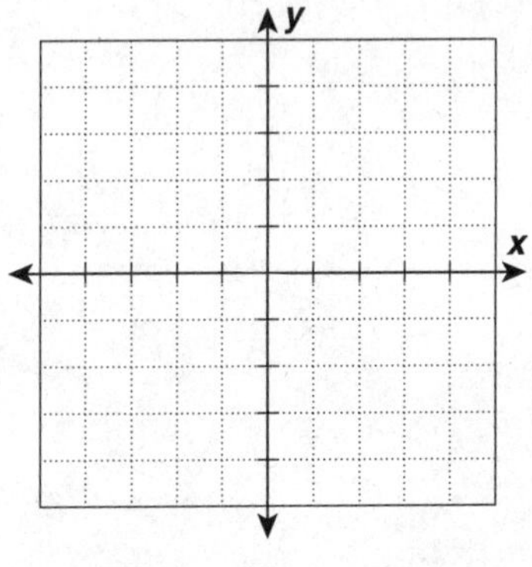

Name ______________________ Date __________ Class __________

LESSON 9-2

Technology Lab

Graph Piecewise Functions continued

12. $h(x)\begin{cases} 2x + 8 & \text{if } x \leq -2 \\ x^2 & \text{if } x > -2 \end{cases}$

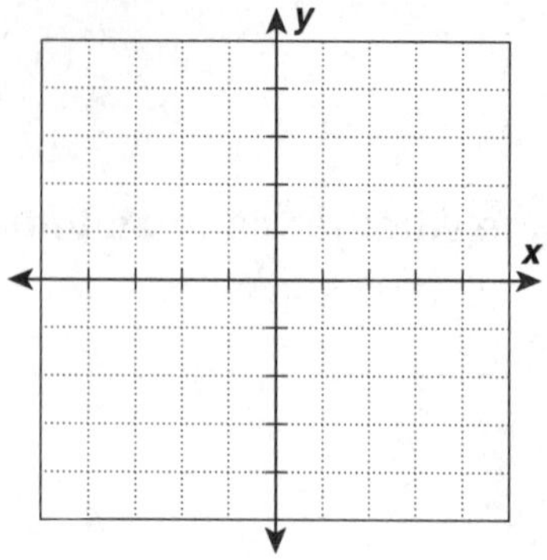

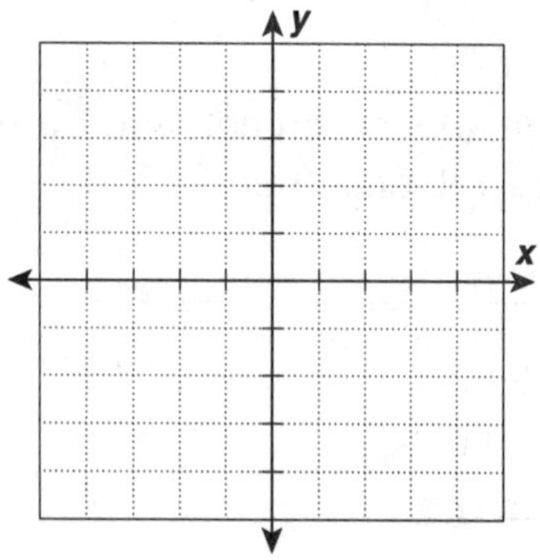

13. $h(x)\begin{cases} -3x & \text{if } x < 1 \\ x - 4 & \text{if } 1 \leq x < 5 \\ -\frac{1}{2}x + \frac{7}{2} & \text{if } x \geq 5 \end{cases}$

14. **Critical Thinking** Explain how multiplying the function by a logical test value visually restricts the domain of the function on your graph.

____

____

15. **Critical Thinking** Explain why you should use dot mode instead of connected mode when graphing piecewise functions.

____

____

Name ______________________ Date __________ Class __________

LESSON 9-6

Technology Lab Recording Sheet (p. 697)

Explore Differences and Ratios

Try This

For each data set, find the first differences, second differences, and ratios of the *y*-values. Identify which, if any, are constant.

1.

x	2	4	6	8	10
y	−20	124	364	700	11.32

First Differences: ______________________

Second Differences: ______________________

Ratios: ______________________

2.

x	4	7	10	13	16	19
y	1.31	2.48	3.65	4.82	5.99	7.16

First Differences: ______________________

Second Differences: ______________________

Ratios: ______________________

3.

x	3	4	5	6	7
y	8.96	35.84	143.36	573.44	2293.8

First Differences: ______________________

Second Differences: ______________________

Ratios: ______________________

4.

x	1.5	3	4.5	6	7.5
y	−9	15	57	117	195

First Differences: ______________________

Second Differences: ______________________

Ratios: ______________________

5. **Critical Thinking** What type of function might be an appropriate model for Problem 4? Justify your response.

____

Name ____________________ Date __________ Class __________

LESSON 9-6

Technology Lab

Modeling Objects Rolling Down a Ramp

Plotting a ball or toy car rolling down a ramp of varying inclines creates a family of curves, which can be modeled by a series of quadratic equations. This activity investigates the values of the coefficients in the quadratic equation, $y = ax^2 + bx + c$.

MATERIALS

TI-84
CBR 2™ motion detector
unit-to-CBR 2™ or I/O unit-to-unit cable
EasyData application
large (9 inch) playground ball
long ramp (at least 2 meters or 6 feet—a lightweight board works well)
protractor to measure angles
books to prop up ramp

Activity 1

Follow these steps to measure the distance of a ball or toy car as it travels down a ramp.

1. Use the protractor to set the ramp at a 15° incline. Lay the CBR 2™ motion detector on the ramp and flip the sensor head so it is perpendicular to the ramp.

   Mark a spot on the ramp 15 centimeters (about six inches) from the CBR 2™ motion detector. Have one student hold the ball at this mark, while a second student holds the calculator and CBR 2™ motion detector.

   Hints: Aim the sensor directly at the ball and make sure that there is nothing in the *clear zone*.

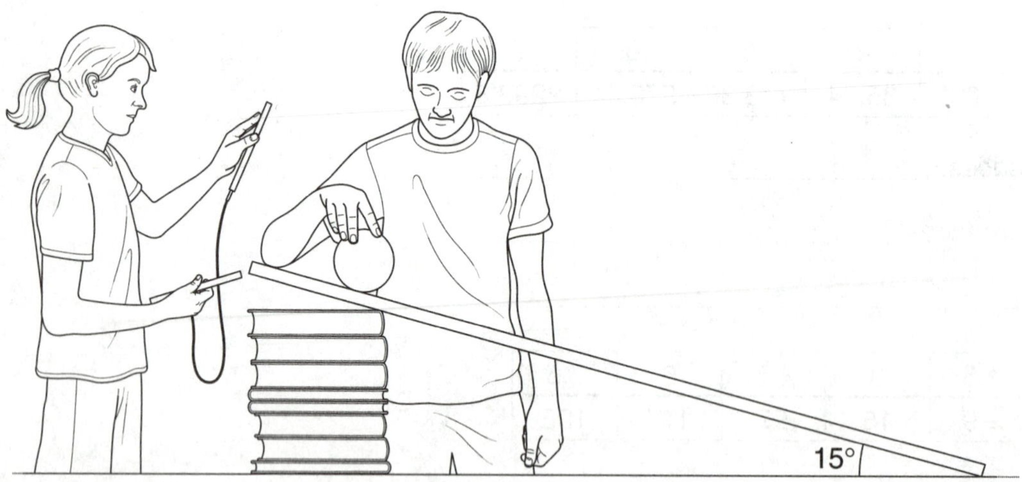

2. Run the EasyData App.
3. To set up the calculator for data collection:
   a. Select Setup (press WINDOW) to open the Setup menu.
   b. Press 2 to select 2: Time Graph to open the Time Graph Settings screen.

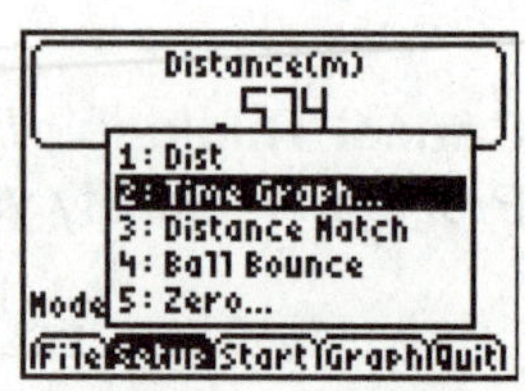

Name ______________________ Date ___________ Class ___________

LESSON **9-6**

Technology Lab

Modeling Objects Rolling Down a Ramp continued

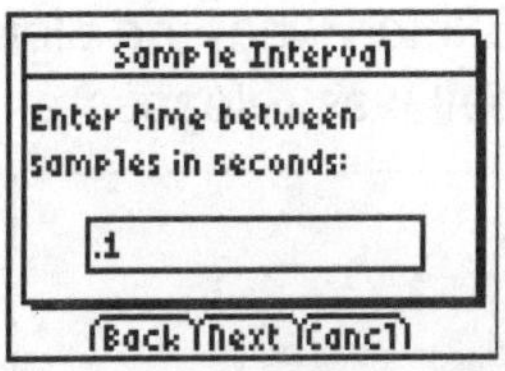

c. Select Edit (press ZOOM) to open the Sample Interval dialog window.

d. Enter 0.1 to set the time between samples in seconds.

e. Select Next (press ZOOM) to advance to the Number of Samples dialog window.

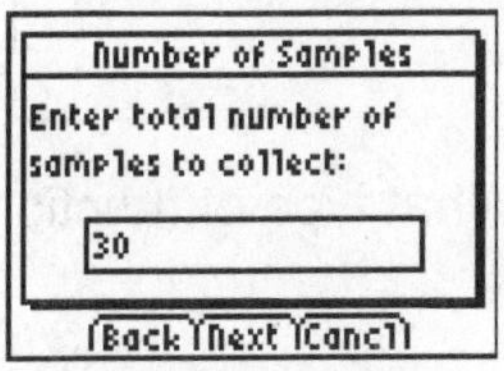

f. Enter 30 to set the number of samples. Data collection will last for 3 seconds.

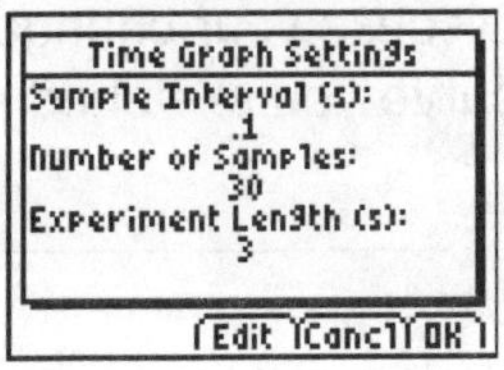

g. Select Next (press ZOOM) to display a summary of the new settings.

h. Select OK (press GRAPH) to return to the main screen.

4. When the settings are correct, choose Start (press ZOOM) to begin sampling.

5. When the clicking begins, release the ball immediately (don't push) and step back.

6. When the clicking stops, the collected data is transferred to the calculator and a plot of distance vs. time is displayed.

Try This

1. Which of these plots do you think best matches the *Distance-Time* plot of a ball rolling down a ramp?

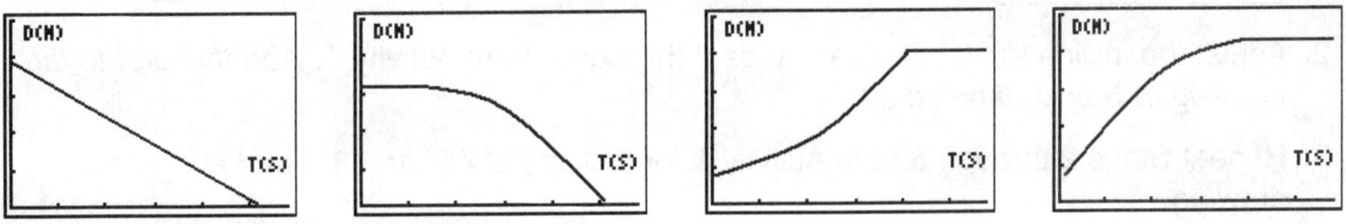

2. What physical property is represented along the *x*-axis? ______________________

What are the units? ______________________

What physical property is represented along the *y*-axis? ______________________

What are the units? ______________________

Name ______________________________________ Date ___________ Class ___________

LESSON 9-6

Technology Lab

Modeling Objects Rolling Down a Ramp continued

3. Sketch what the plot really looks like. Label the axis. Label the plot at the points when the ball was released and when it reached the end of the ramp.

4. What type of function does this plot, between the two points you identified, represent?

____

5. Discuss your change in understanding between the graph you chose in question 1 and the curve you sketched in question 3.

____

____

Activity 2

Examine what happens for differing inclines.

1. Predict what will happen if the incline increases. Create a sketch of the predicted plot.

2. Adjust the incline to 30°. Repeat Steps 2 through 6 from Activity 1. *Add this plot to the drawing in Step 1*, labeled 30°.

3. Repeat Steps 2 through 6 from Activity 1 for inclines of 45° and 60° and add to the drawing.

4. Sketch and label the plots for 0° and 90°.

Name ________________________________ Date ___________ Class ___________

LESSON 9-6

Technology Lab

Modeling Objects Rolling Down a Ramp continued

Try This

Adjust the time values so that $x = 0$ for the initial height (the time at which the ball was released. You can do this manually by subtracting the *x* value for the first point from all the points on your plot, or you can enter L1(1)→A:L1→L1.

6. Calculate the values for *a*, *b*, and *c* for the family of curves in the form $y = ax^2 + bx + c$ at 0°, 15°, 30°, 45°, 60°, 90°.

____

____

____

7. What are the minimum and maximum values for *a*? Why?

____

____

____

8. Write an expression describing the mathematical relationship between *a* and the angle of inclination.

____

Name ______________________ Date __________ Class __________

LESSON **10-3**

Technology Lab
Graphing Conic Sections

Use with Lesson 10-3

A commercial builder is developing a new mall and plans to add a large brick patio in the shape of an ellipse. He plans to base his design of the ellipse on the equation $\frac{x^2}{676} + \frac{y^2}{225} = 1$. or on the equation $\frac{x^2}{100} + \frac{y^2}{169} = 1$.

Activity 1

Use your graphing calculator to graph the ellipse.

Step 1 Press APPS and scroll down to **6: Conics** and press ENTER.

Scroll down to **2:ELLIPSE**, press ENTER and then press **1**.

Step 2 You now have to input that **A=26, B=15, H=0, and K=0.**

Step 3 Press GRAPH.

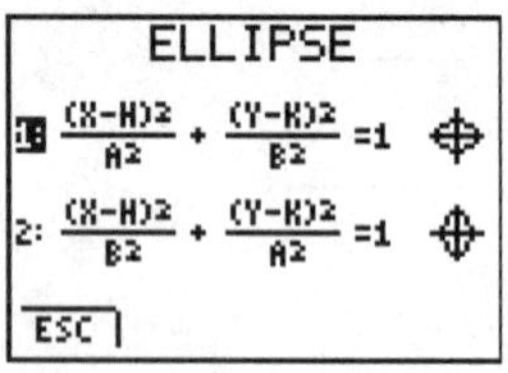

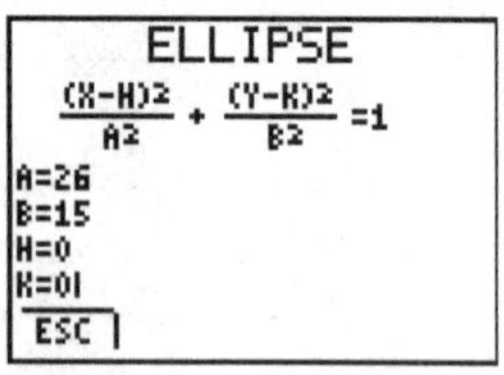

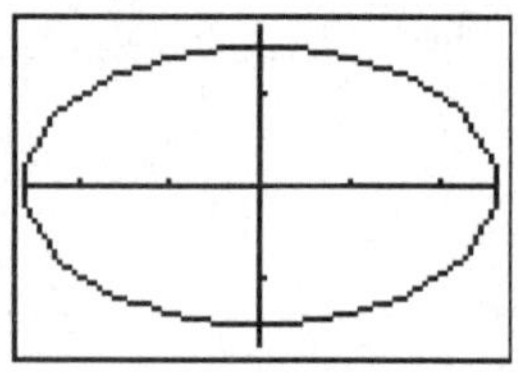

Try This

Graph the ellipse given by the equation.

1. $\frac{x^2}{169} + \frac{y^2}{100} = 1$

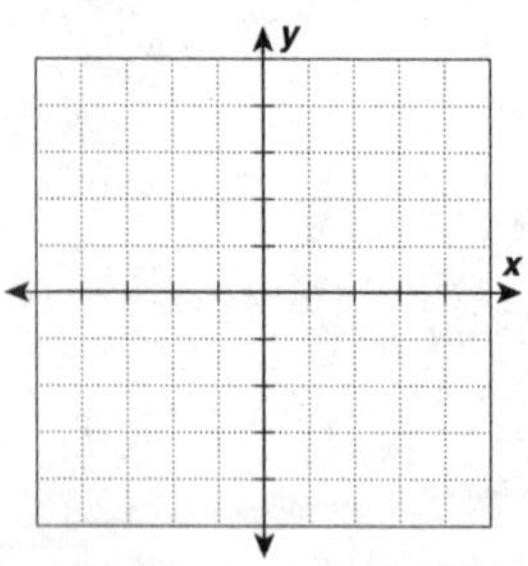

2. $\frac{(x-2)^2}{196} + \frac{(y-4)^2}{144} = 1$

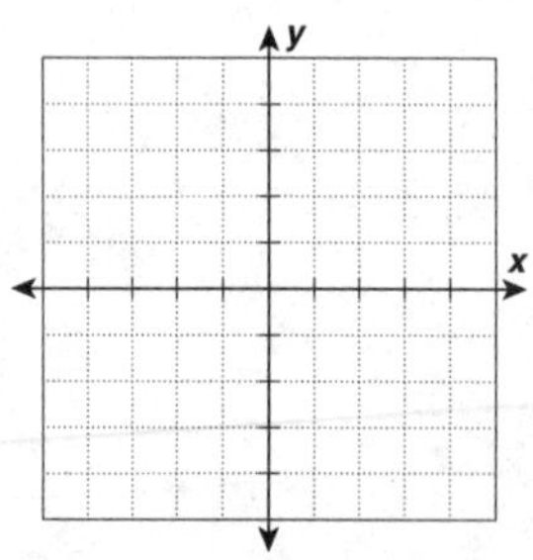

3. Critical Thinking If the commercial builder based his design on the equation of the ellipse $\frac{x^2}{100} + \frac{y^2}{169} = 1$, how would Step 1 described above differ? Why?

____

____

____

Name ______________________ Date __________ Class __________

LESSON **10-3**

Technology Lab

Ellipse continued

Activity 2

At a local model rocket competition officials keep record of the results of the best launches. Last years winner based his design on the elliptical trajectory equation $\frac{(x-22)^2}{324} + \frac{y^2}{1764} = 1$. Use your graphing calculator to graph this equation.

Step 1 Begin by pressing APPS scroll down to **6:Conics** and press ENTER.

Scroll down to **2:ELIIPSE** and press ENTER. Now press **2** to choose the correct type of ellipse.

Step 2 Input the values **A=42, B=18, H=22, K=0.**
(Note: $\sqrt{1764} = 42$; $\sqrt{324} = 18$)

Step 3 Press GRAPH.

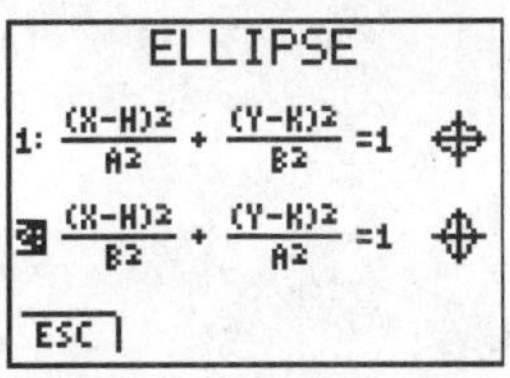

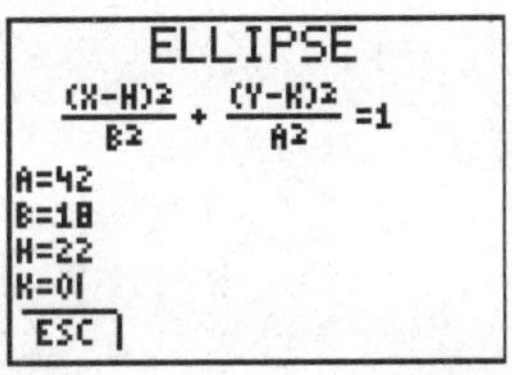

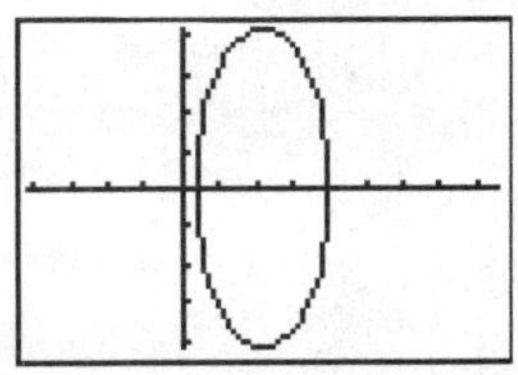

Try This

A designer recommends an elliptical-shaped design for a window. The designer bases his design on the equation $\frac{x^2}{25} + \frac{y^2}{49} = 1$. He then changes his mind and uses a new equation with an *a*-value of 14 and a *b*-value of 10.

1. What is the equation of the new window?

____

2. Graph both equations on your calculator.

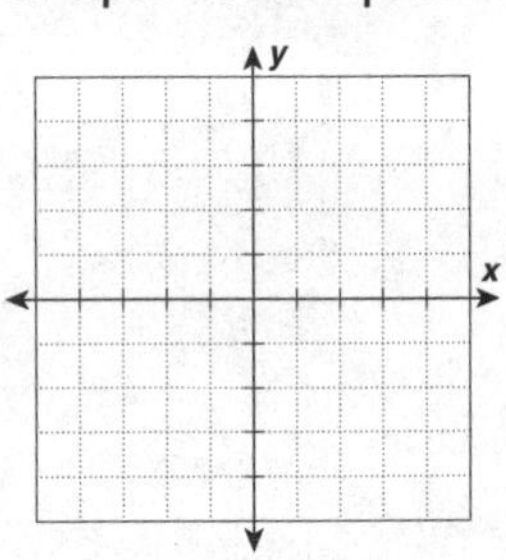

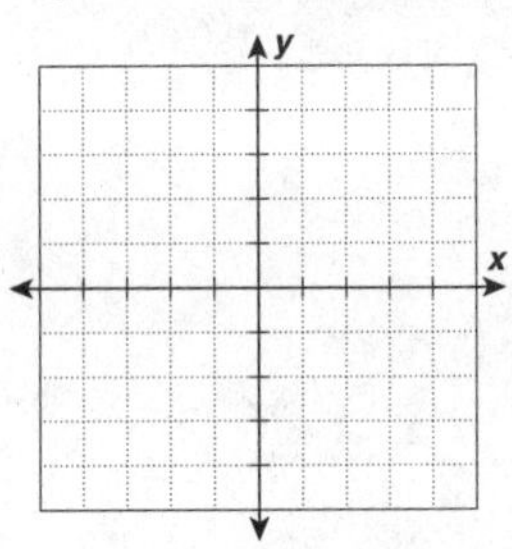

3. How do the graphs differ? How do they compare?

____

____

____

Name ______________________________ Date __________ Class __________

LESSON 10-6

Technology Lab Recording Sheet (p. 767)

Conic Section Art

Try This

1. Create your own picture by using the graphs of conic sections. Use at least four conic sections. You may also use lines if necessary.

2. Trade equations with a classmate, and attempt to re-create his or her picture by using only the equations.

Name ______________________________ Date __________ Class __________

LESSON **11-2**

Technology Lab Recording Sheet (p. 810)

Explore Simulations

Try This

Use a simulation to find each experimental probability.

1. An energy drink game advertises a 25% chance of winning with each bottle cap. Find the experimental probability that a 6-pack will contain at least 3 winners.

____

2. In a game with a 40% chance of winning, your friend challenges you to win 4 times in a row. Find the experimental probability of this happening in the next 4 games.

____

3. **Critical Thinking** How would you design a simulation to find the probability that a baseball player with a .285 batting average will get a hit in 5 of his next 10 at bats?

____

Name ______________________________ Date ____________ Class ____________

LESSON **11-5**

Technology Lab
Measures of Central Tendency and Dispersions

Use with Lesson 11-5

Activity

We Zapem Pest Control wants to predict how many opossums they will catch next year. The table shows the number of opossums they have caught so far this year. Use your graphing calculator to find the mean number of opossums caught and the standard deviation.

Month	JAN	FEB	MAR	APR	MAY	JUN	JUL	AUG	SEP	OCT	NOV	DEC
Number of Opossums Caught	11	13	15	17	9	45	12	14	15	14	13	15

Step 1 Press STAT and then press 1.

Step 2 Input the monthly numbers by entering **11** for January and then ENTER Then enter **13** for February and press ENTER. Keep entering the values from the table. When you entered the last number press STAT and then press the right arrow →.

Step 3 In the **CALC** menu press **1** then ENTER.

```
1-Var Stats
 x̄=16.08333333
 Σx=193
 Σx²=4065
 Sx=9.346446416
 σx=8.94854116
↓n=12
```

Because the standard deviation is so high, 8.95, identify and remove any outliers from the data set and repeat the above steps.

The outliner you need to remove is **45** June. Input the monthly numbers again but leave out June so that you only have 11 months.

```
1-Var Stats
 x̄=13.45454545
 Σx=148
 Σx²=2040
 Sx=2.207425485
 σx=2.104697619
↓n=11
```

Removing the outlier in the data set causes the mean to change from 16.08 to 13.45 and the standard deviation to change from 8.9 to 2.1. This model is a better indicator of the number of opossums the pest control company is likely to expect.

Name ______________________ Date __________ Class __________

LESSON 11-5

Technology Lab

Measures of Central Tendency and Dispersions continued

Try this

A local grocer is looking for a supplier of natural foods. He has found many suppliers but has had difficulty deciding who to pick to supply eggs to his store. One of the egg suppliers, Jerry's Farm, supplies a low but constant supply of cases of eggs. The second supplier, Mega Industries, can supply larger quantities of eggs, which the grocer likes, but the amount of eggs supplied varies. Use the following table and your graphing calculator to determine the mean and the standard deviation from the weekly cases of egg supplied from both egg producers.

Weeks	1	2	3	4	5	6	7	8	9	10	11
Jerry' s Farm	8	1	7	8	10	9	11	9	7	8	6
Mega Industries	14	12	3	17	6	29	12	7	15	4	13

1. What is the mean and standard deviation from Jerry's egg production?

____

2. What is the mean and standard deviation from Mega's egg production?

____

3. **Critical Thinking** The grocer has found that he needs about 9 cases of eggs a week. Which supplier of eggs will be closest to meeting his goal? Why? Remember to take out the outliner for an accurate choice.

____

____

____

Name ______________________ Date __________ Class __________

LESSON 12-2

Technology Lab Recording Sheet (p. 878)

Evaluate Sequences and Series

Try This

Find the first 8 terms of each sequence. Then find the sum of those 8 terms.

1. $a_n = 2n^2 - 5$

____

2. $a_n = \frac{1}{4}(2)^{n-1}$

____

3. $a_n = 0.3n + 1.6$

____

4. $a_n = 20n$

____

5. $a_n = n^3 - 2n$

____

6. $a_n = 0.1(5)^n$

____

7. Critical Thinking Find the next 5 terms of the sequence 200, 182, 164, 146, 128, ...

____

Then find the sum of those 5 terms. __________

Name ____ Date ____________ Class ____________

LESSON **12-4**

Technology Lab

Finding the Sum of a Geometric Series

Use with Lesson 12-4

Activity

Use a calculator to find the sum of a geometric series.

Follow this procedure to find the sum of the series $\sum_{k-1}^{8} -4(30)^{k-1}$.

Begin by pressing [2nd] and then press [STAT].

Scroll right to the **MATH** menu.

Press **5** for **sum (**.

Now press [2nd] and [STAT] and scroll right to the **OPS** menu.

Press **5** for **seq (**.

Press [(−)] and then **4** [(] **3** [)].

(Do not substitute [−] for [(−)].)

Now press [^] [(] [X,T,θ,n] [−] and then 1 [)] [,] [X,T,θ,n] [,].

Now press 1 [,] 8 [,] 1 [)] [)] [ENTER].

Your answer should read **−13120**.

Try This

Use a calculator to find the sum of the following geometric series.

1. $\sum_{k-1}^{4} 2(2)^{k-1}$

____

2. $\sum_{k-1}^{5} \left(\frac{1}{2}\right)^{k-1}$

____

3. $\sum_{k-1}^{11} (3)^{k-1}$

____

Name ______________________ Date ____________ Class ____________

LESSON 13-3

Technology Lab Recording Sheet (p. 942)

Explore the Unit Circle

Try This

Use the unit circle on your graphing calculator to determine the sine and cosine of each angle.

1. $\theta = 150°$ ______________________

2. $\theta = 244°$ ______________________

3. $\theta = 90°$ ______________________

4. **Make a Conjecture** How can you verify that the unit circle displayed on your graphing calculator has a radius of 1 unit?

____

____

____

____

____

5. **Make a Conjecture** Use the graph of the unit circle to explain why the sine function is negative for an angle θ in standard position if the angle's terminal side lies in Quadrants III or IV.

____

____

____

____

____

Name ______________________________ Date ____________ Class ____________

LESSON 13-6

Technology Lab

Using the Law of Cosines

Use with Lesson 13-6

Activity

Use a graphing calculator to solve problems involving the Law of Cosines.

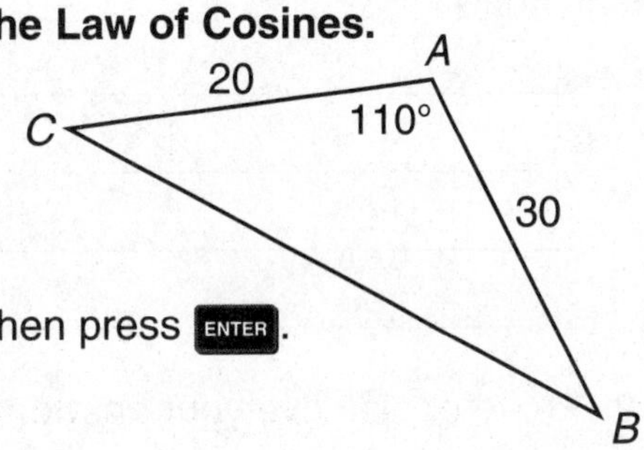

Find a if $b = 20$, $c = 30$, and $m\angle A = 110°$. Round your answer to the nearest tenth.

To begin, you need to set your calculator to degree mode.

Press [MODE], use the arrow keys to move to DEGREE, and then press [ENTER].

Press [2nd] [MODE].

By the Law of Cosines, $a^2 = b^2 + c^2 - 2bc \cos A$.

$a = \sqrt{b^2 + c^2 - 2bc \cos A}$ Solve for a.

$a = \sqrt{20^2 + 30^2 - 2(20)(30) \cos 110}$ Substitute known values.

Enter the equation into your calculator by following these steps.

Press [2nd] [x^2] 20 [x^2] [+] 30 [x^2] [−] 2 [(] 20 [)] [(] 30 [)] [COS] 110 [)]

To the nearest tenth, $a = 41.4$.

Try This

Use the Law of Cosines to find the indicated side length of $\triangle ABC$. Round your answers to the nearest tenth.

1. Find a if $b = 50$, $c = 42$, and $m\angle A = 126°$. ____________
2. Find b if $a = 64$, $c = 48$, and $m\angle B = 119°$. ____________
3. Find c if $a = 22$, $b = 16$, and $m\angle C = 168°$. ____________
4. The ferryboat, Enola Gray, has left shore on a site seeing trip. A passenger realizes he mistakenly boarded the ferryboat named Endeavor. The Enola Gray travels at a speed of 9 nautical miles per hour (9 knots), The Endeavor travels at a speed of 4 nautical miles per hour (4 knots). How long will it take the Endeavor to catch up with the Enola Gay if the angle between each of their course is 99° and they have been traveling for 3 hours?

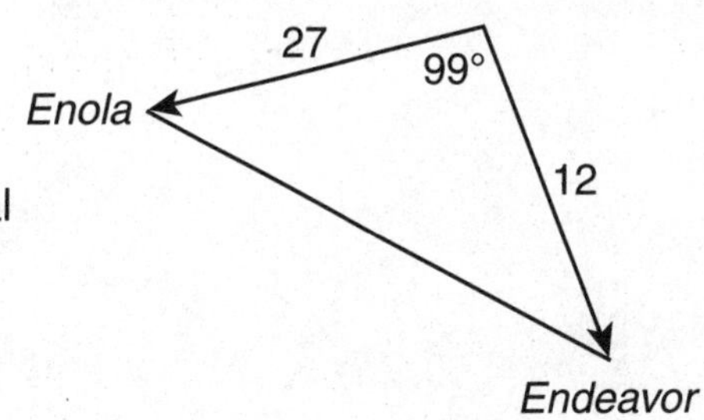

____

Name ______________________ Date ____________ Class ____________

LESSON 14-3

Technology Lab Recording Sheet (p. 1006)

Graph Trigonometric Identities

Try This

1. **Make a Conjecture** Determine whether $\sec x - \tan x \sin x = \cos x$ is a possible identity.

____

____

____

____

2. Prove or disprove your answer to Problem 1 by using algebra.

____

____

____

____

3. **Make a Conjecture** Determine whether $\frac{1 + \tan x}{1 + \cot x} = \tan x$ is a possible identity.

____

____

____

____

4. Prove or disprove your answer to Problem 3 by using algebra.

____

____

____

____

____

Name ______________________________ Date ___________ Class ___________

LESSON 14-3

Technology Lab

Graphing to Check for Equivalent Functions Expressions

Use with Lesson 14-3

You can use a graphing calculator to check for equivalent expressions.

Activity

The expression $\frac{\sin^2\theta}{1-\cos\theta}$ rewritten in terms of $\cos\theta$, can be simplified to $1+\cos\theta$. Check whether these expressions are equivalent.

First set the calculator to radian mode by pressing [MODE] and using the arrow keys to move to RADIAN. Then press enter. Press [2nd] [MODE].

Enter the first equation at Y1=:

Begin by pressing [Y=] and then press [SIN] [X,T,θ,n] [)] [x^2].

[÷] [(] 1 [−] [COS] [(] [X,T,θ,n] [)] [)].

Enter the second equation at Y2=:

At Y2 press **1** [+] [COS] [(] [X,T,θ,n] [)].

Press [GRAPH].

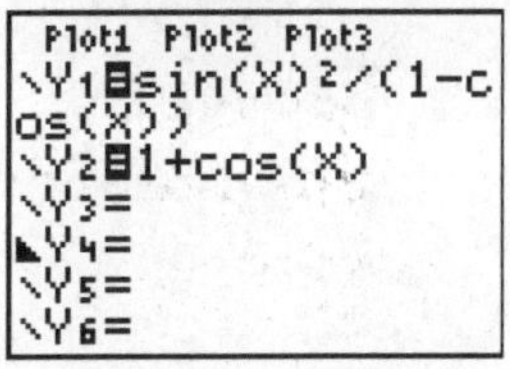

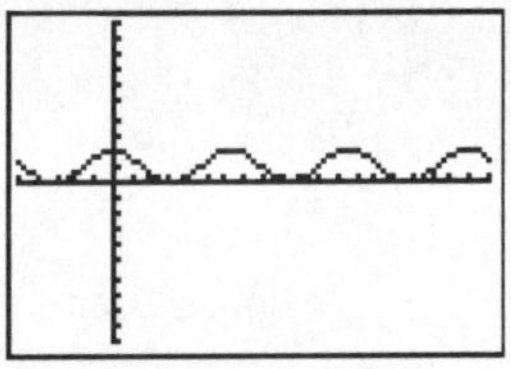

The graphs of each function are the same, so the expressions are equivalent.

Try This

1. Write $\frac{1}{\sec^2\theta}$ in terms of $\sin\theta$ and simplify. Check if your answer's graph appears to be the same as $\frac{1}{\sec^2\theta}$.

   ____

2. Write $\sec\theta - \tan\theta\sin\theta$ in terms of $\sin\theta$ and check if its graph appears to be the same as $y = \cos\theta$.

   ____

Answer Key

CHAPTER 1

Lesson 1-5

Try This

1. 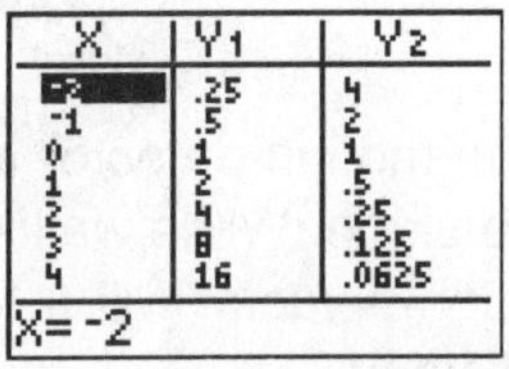

X	Y1	Y2
-2	.25	4
-1	.5	2
0	1	1
1	2	.5
2	4	.25
3	8	.125
4	16	.0625

X= -2

2. 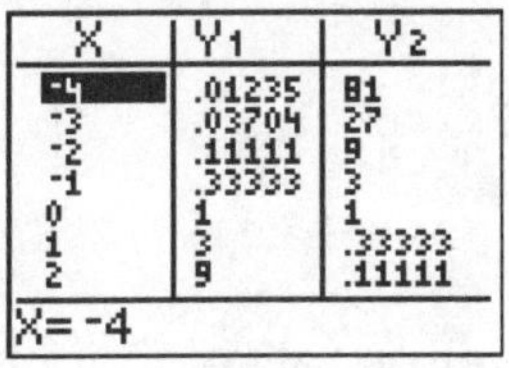

X	Y1	Y2
-4	.01235	81
-3	.03704	27
-2	.11111	9
-1	.33333	3
0	1	1
1	3	.33333
2	9	.11111

X= -4

3. 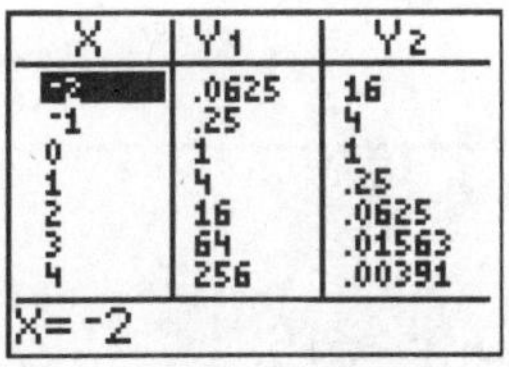

X	Y1	Y2
-2	.0625	16
-1	.25	4
0	1	1
1	4	.25
2	16	.0625
3	64	.01563
4	256	.00391

X= -2

4. 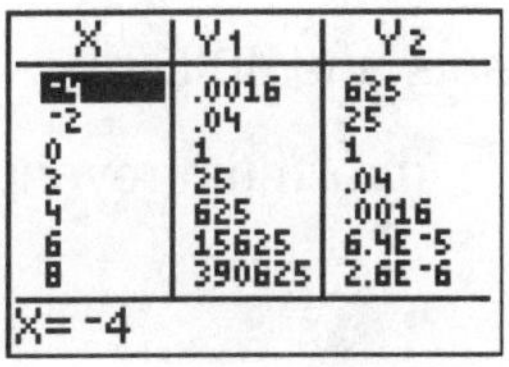

X	Y1	Y2
-4	.0016	625
-2	.04	25
0	1	1
2	25	.04
4	625	.0016
6	15625	6.4E-5
8	390625	2.6E-6

X= -4

5. 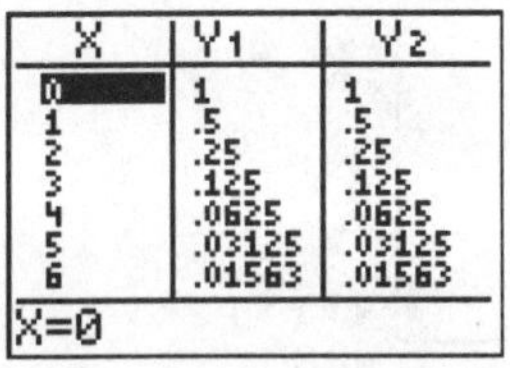

X	Y1	Y2
0	1	1
1	.5	.5
2	.25	.25
3	.125	.125
4	.0625	.0625
5	.03125	.03125
6	.01563	.01563

X=0

6. The product is always 1.
7. $\{x \mid x < 0\}$

Lesson 1-9

Try This

1. 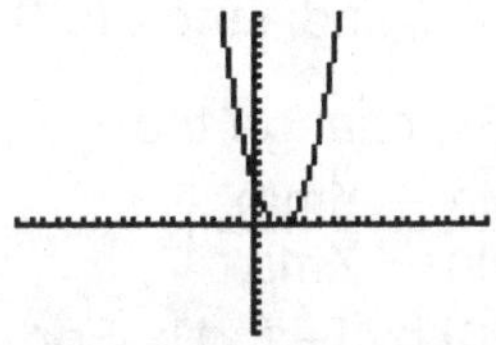

The graph of $y = (x - 2)^2$ is horizontally translated 2 units to the right from the parent function.

2. $y = x^2 + 8$
3. The graph is translated 4 units down.

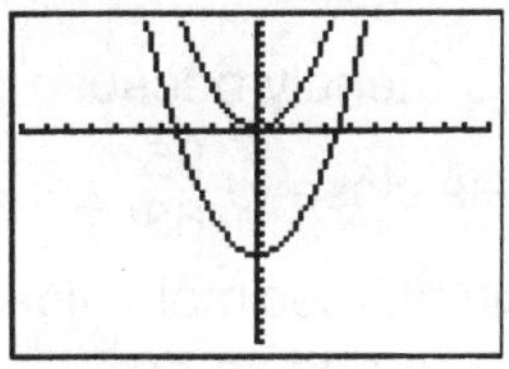

CHAPTER 2

Lesson 2-3

Activity 1

1. x: $[-20, 20]$; y: $[-30, 10]$

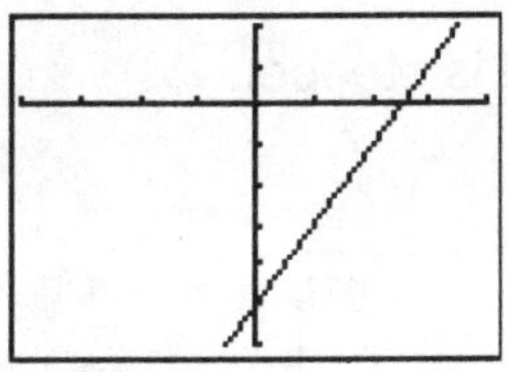

2. x: $[-40, 10]$; y: $[-60, 10]$

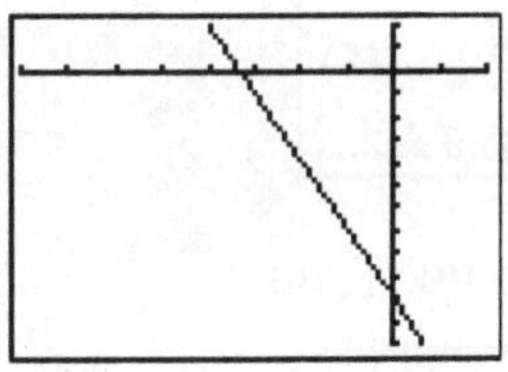

3. x: $[-40, 10]$; y: $[-30, 30]$

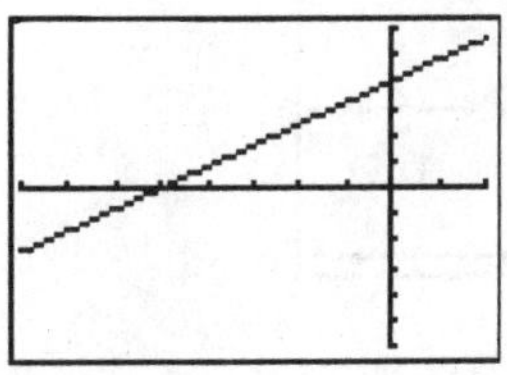

4. Zoom out, use the **ZInteger** window, or decrease **Xmin** and **Ymin**.

5. Change **Ymin** to −0.1 and **Ymax** to 0.1.

6. For a steep $y = 0.5x$, change the window so that $\frac{\text{Ymin} - \text{Ymax}}{\text{Xmin} - \text{Xmax}}$ is small, e.g., [−20, 20] by [−1, 1]; For a flat $y = 10x$, change the window so that $\frac{\text{Ymin} - \text{Ymax}}{\text{Xmin} - \text{Xmax}}$ is large, e.g., [−1, 1] by [−50, 50].

Activity 2

1. **Xmax − Xmin** = 18.8; **Ymax − Ymin** = 12.4; The window is friendly because $\frac{188}{124} \approx 1.52$, which is close to $\frac{95}{63}$, and tracing results in simple decimal values for x, such as 1.1, 1.3, and so on.

2. Graph in [−47, 47] by [−55, 6].

3. Possible answers:

   a. Use a square window.

   b. **ZDecimal**

Lesson 2-7

1. The second graph is steeper.

2. Distance vs. Time

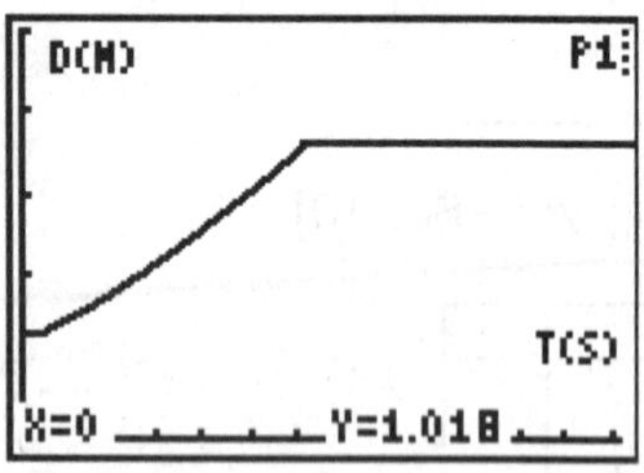

Matching Distance vs. Time

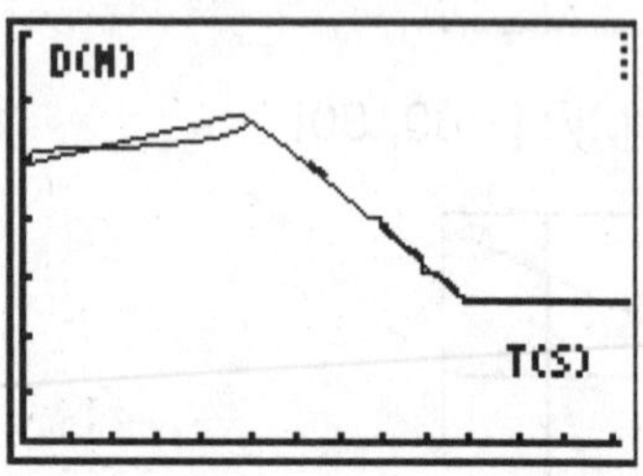

The slope of the portion of the graph corresponding to movement is greater for the faster trial.

Results will probably vary between groups as they may walk at different rates.

Walking towards the motion detector will produce a negative slope. While walking away from the motion detector will produce a positive slope.

Lesson 2-9

1. {−1, 3}

2. {−3}

3. The left side is never negative; there is no intersection point.

CHAPTER 3

Lesson 3-4

1. $x + y \leq 80$, $x \geq 20$, $y \geq 15$;
   $P = 7.5x + 8.25y$

2. (40, 40): 40 parakeets and 40 Doves

3. $P = 7.5x + 8.25y$; a maximum revenue of $630

Lesson 3-6

1. a. 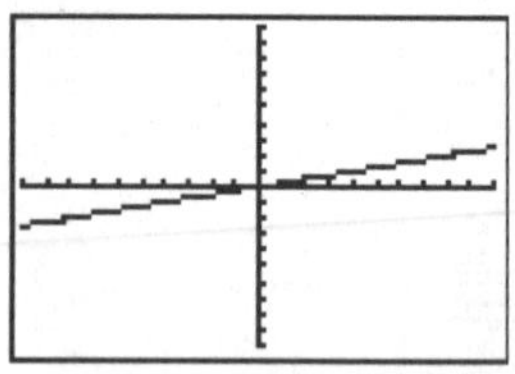

   $m = \frac{1}{4}$

   b. Possible answer: The slope is the quotient of the $y1$ coefficient and the $x1$ coefficient.

2. a. 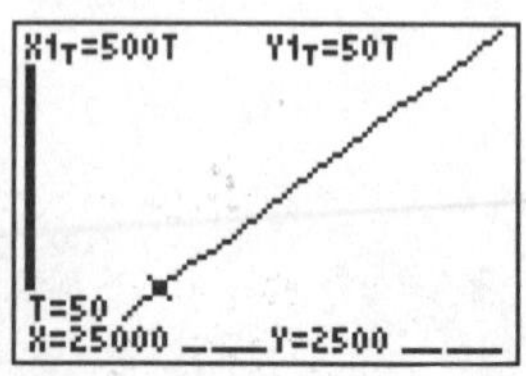

   (25,000, 2500)

b. 100 s

3. $\begin{cases} x = t \\ y = t \end{cases}$

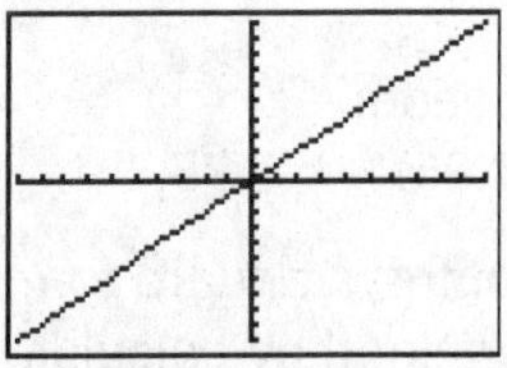

CHAPTER 4

Lesson 4-5

1. $(-1.769, 3.692)$
2. Possible answer: Multiply the coefficient matrix by the solution matrix and check whether this product is equal to the constant matrix. For example, enter **=B2*D10+C2*D11** in cell 13D and **=B3*D10+C3*D11** in cell 13E. Check whether the entries in cells 13D and 13E are equal to the entries in cells 10D and 10E.
3. Possible answer: For both an inconsistent system and a dependent system, $D = 0$ and the entries in the inverse of A and in the solution matrix are undefined. Replace a column of coefficients of A by the constants and calculate the determinant. Then put the original values back and replace the other column with constants. If either determinant is 0, the system is dependent, otherwise it is inconsistent.

Lesson 4-6

1. Child \$6.00, Adult \$7.50, Senior \$6.75
2. \$8.00
3. \$75.00

CHAPTER 5

Lesson 5-1

1. The graph of $g(x) = (x - 4)^2$ is the graph of f translated 4 units right.
2. The graph of $g(x) = (x + 3)^2$ is the graph of f translated 3 units left.
3. The graph of $g(x) = -x^2$ is the graph of f reflected across the x-axis.
4. The parameter h in the function $g(x) = (x - h)^2$ has the effect of translating the parent function $f(x) = -x^2$ h units right if h is positive and h units left if h is negative.
5. The parameter a in the function $g(x) = ax^2$ has the effect of vertically stretching the parent function $f(x) = x^2$ if a is greater than 1 and vertically compressing the parent function if a is between 0 and 1.

Lesson 5-3

1.

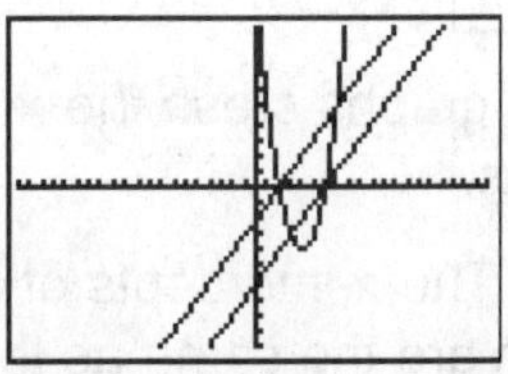

2, 6; $x = 4$

2.

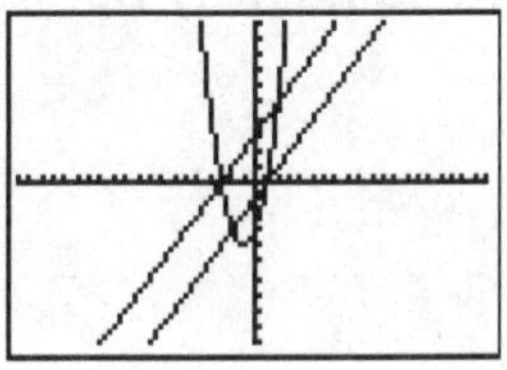

-3, 1; $x = -1$

3.

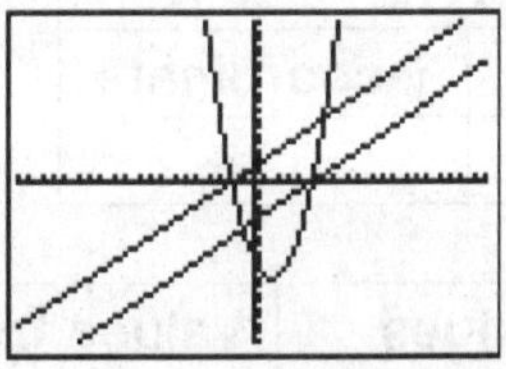

-3, 1; $x = -1$

4.

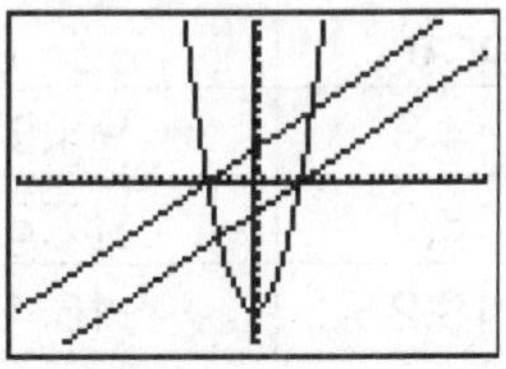

-3, 1; $x = -1$

5. 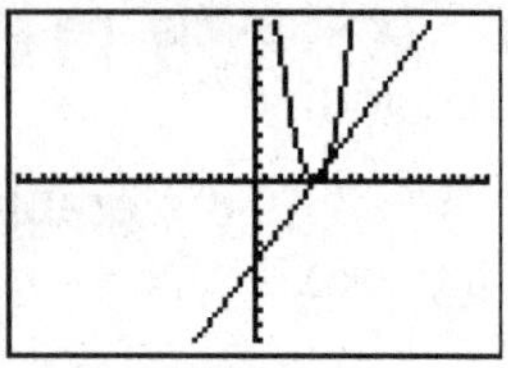

   5; $x = 5$

6. 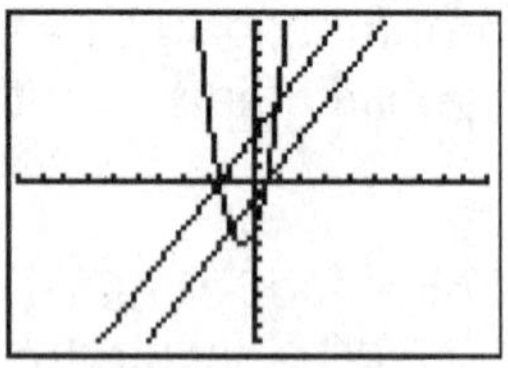

   $\frac{1}{2}, -\frac{3}{2}; x = -\frac{1}{2}$

7. 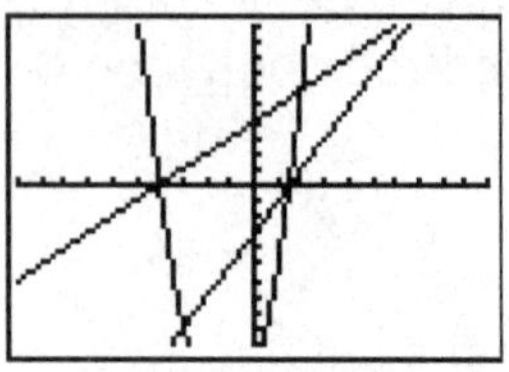

   Yes, because the graphs cross the *x*-axis at the same points.

8. Possible answer: The *x*-intercepts of a quadratic function are the same as the *x*-intercepts of its linear factors. The axis of symmetry is located halfway between the *x*-intercepts.

Lesson 5-8

Data Tables

Sample data; actual data may vary.

Vertex	
***x*-coordinate**	***y*-coordinate**
1.95	0.78

Parameters	Values calculated from vertex form	Values from regression
a	−4.8	−4.6
b	18.9	17.8
c	−19.2	−16.4

Try This

1. The graph we are using is vertical distance versus time, not vertical distance versus horizontal distance. That is, the horizontal axis is not horizontal distance, so the appearance of the graph has nothing to do with a side-ways moving ball.
2. Vertex model equation: $y = -4.85\,(x - 1.95)^2 + 0.78$.
3. The parameters of the standard form quadratic as determined by calculator regression and by the vertex fit are similar.
4. The magnitude of *a* determines how sharply curved the parabola is, while the sign of *a* determines whether the parabola is open upward (positive *a*) or downward (negative *a*).
5. Since the vertex of the new parabola would be to the right of the one originally used, the time value *h* would be larger. The *y* coordinate of the vertex would be smaller than before, as the ball does not bounce as high each time.

CHAPTER 6

Lesson 6-7 pp. 24–25

1. **a.** Possible answer: They are all polynomials of degree 3 with positive leading coefficients.

   b. Possible answer: The functions have the same end behavior, rising for positive *x*-values and falling for negative *x*-values.

   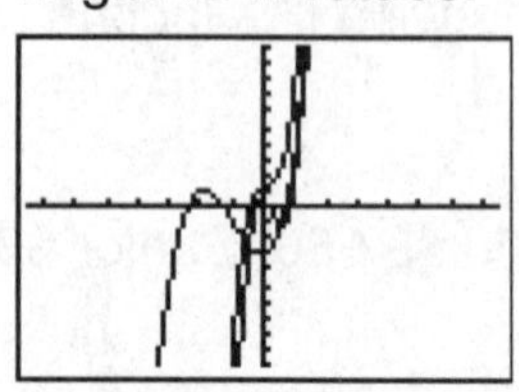

   c. Possible answer: The end behavior will be the same as $g(x)$, $h(x)$, and $k(x)$, rising for positive *x*-values and falling for negative *x*-values.

2. **a.** Possible answer: They are all polynomials of degree 3 with negative leading coefficients.

b. Possible answer: The functions have the same end behavior, rising for negative *x*-values and falling for positive *x*-values.

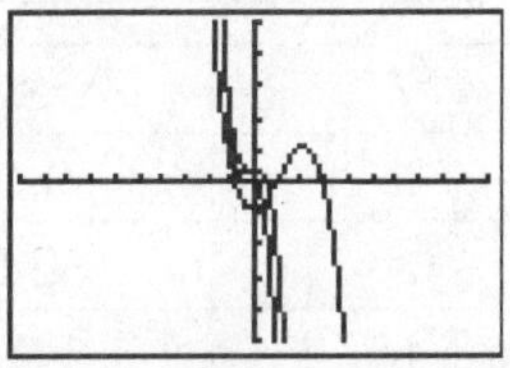

c. Possible answer: The end behavior will be the same as *a*(*x*), *b*(*x*), and *c*(*x*), rising for negative *x*-values and falling for positive *x*-values.

3. a. Possible answer: They are all polynomials of degree 4 with positive leading coefficients.

b. Possible answer: The functions have the same end behavior, rising for both positive and negative *x*-values.

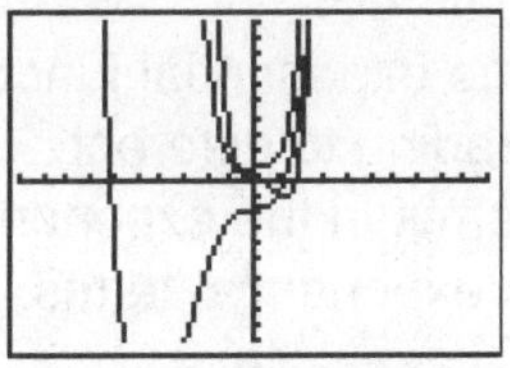

c. Possible answer: The end behavior will be the same as *p*(*x*), *r*(*x*), and *s*(*x*), rising for both positive and negative *x*-values.

4. Possible answer: The degree of the polynomial and sign of the leading coefficient affect end behavior.

Lesson 6-7 pp. 26–27

1. minimum (1, 6); maximum (−1, 18)

2. minimum (1.4142, −1.656); maximum (−1.4142, 9.656)

CHAPTER 7

Lesson 7-2

1. D: $\mathbb{R}$; R: $\{x \mid x \geq 0\}$; inverse: D: $\{x \mid x \geq 0\}$; R: $\mathbb{R}$; No; it fails the vertical line test for $\{x \mid x > 0\}$.

2. D: $\mathbb{R}$; R: $\mathbb{R}$ for both; Yes; for any input, there is at most 1 output.

3. The domain of a function is the range of the inverse, and the range of a function is the domain of the inverse.

4. They reflect across $f(x) = x$.

Lesson 7-6

1. Rate
A3
0.035

A4
0.04

A5
0.045

⋮

⋮

A19
0.115

A20
0.122.

2. Doubling Time
B3
20.14879

B4
17.67299

B5
15.7473

⋮

⋮

B19
6.367654

B20
6.116255

3. Product

C3
0.703493

C4
0.705208

C5
0.70692

⋮
⋮

C19
0.73228

C20
0.733951

4. 8.5%

5. The product of the interest rate written as a percent and the doubling time is about 72.

6. $d = \frac{0.72}{r}$

Lesson 7-8

Sample Results

Actual data will vary.

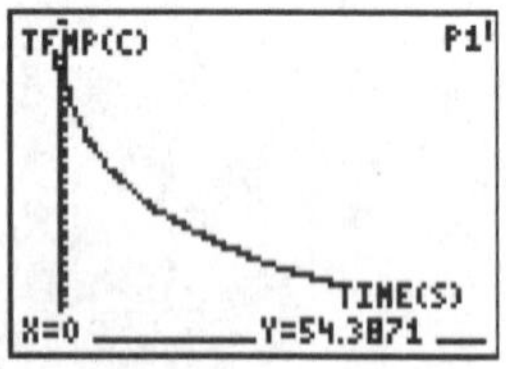

Uncorrected temperature versus time data

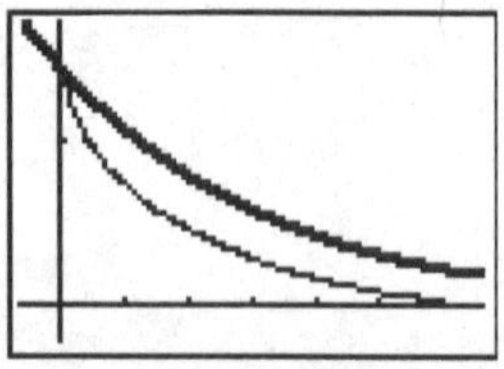

Data and model with first guess for k (0.01)

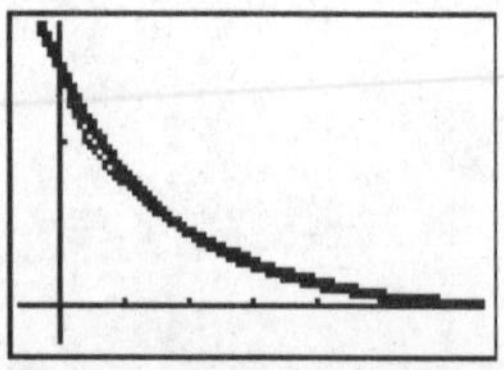

Data and regression line

Data Table

Sample data; actual data will vary.

Minimum Temperature (°C)	24

$y = Te^{-kt}$	T	30
	k	0.02
$y = ab^x$	a	38.03
	b	0.977

1. Yes, the data is consistent with a decreasing exponential, as the graph is always decreasing, but it never crosses the horizontal axis.
2. Model equation is $y = 30e^{-0.02x}$
3. The regression line fits the data well. Both the regression and the model are exponential functions, so they can have the same shape.
4. The bases of the exponential functions are different, leading to different multiplicative terms in the exponent. Comparing the exponential terms, we have $b^x = e^{-kx}$, or $-\ln(b) = k$.
5. When $t = 0$, we have $e^0 = 1$.
6. When t is large, $e^{-\text{large}}$ approaches zero, so the temperature difference approaches zero.
7. A decreasing k would correspond to a longer time to cool. To make the probe take longer to cool, wrap it in insulation.
8. $1 = 30e^{-0.02t}$, or $t = 170$ seconds, using the model.
9. No, it takes more than half as long. Using the model expression, we see that the time to 1°C temperature difference is ln(starting temp difference)/0.02. If the starting temp difference is 15°C, then it takes 135 seconds to reach a 1°C difference.

CHAPTER 8

Lesson 8-1

Sample Results

Actual data will vary.

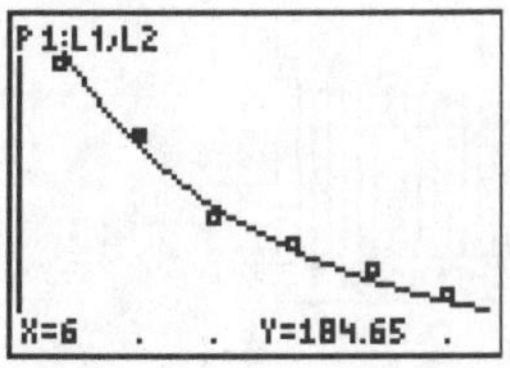

Sample data with inverse-square model

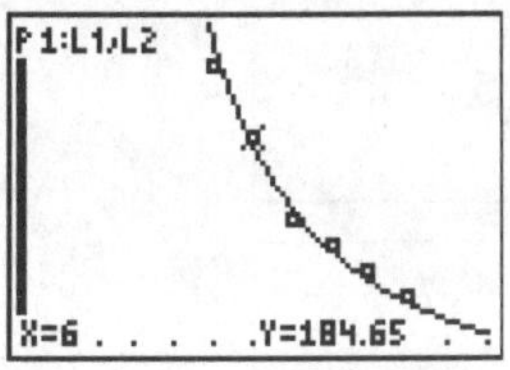

Same sample data replotted over a wider range

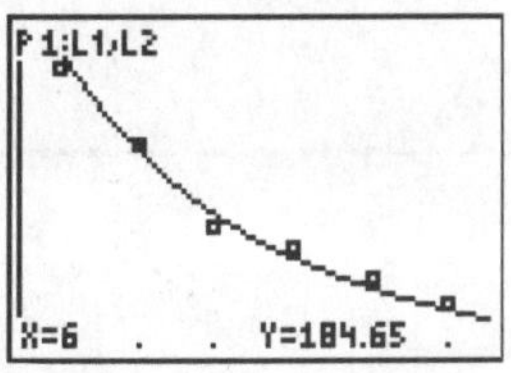

Sample data with power law model

```
PwrReg
 y=a*x^b
 a=4059.647321
 b=-1.737988357
```

Power law fit parameters

Data Table

Sample data; actual data may vary.

Left-most point x value x_L	5
Left-most point y value y_L	244
Initial model parameter C	6100
Optimized inverse-square model $y = C/x^2$	$y = 6300/x^2$
Power law model $y = ax^b$	$y = 4100x^{-1.8}$

Answers to Questions

1. The intensity of the light appears to get smaller with increasing distance. The relationship could be an inverse relationship, or it could be something else.
2. Yes, the data are consistent with a decreasing function such as an inverse function, as the graph is always decreasing but it never crosses the horizontal axis.
3. The inverse-square model fits the data very well.
4. The fit of the general power law looks very similar to the inverse-square fit. The power law fit has to be at least as good as the inverse-square fit, since the power law fit includes the possibility of the inverse-square fit with the exponent taking a value of -2. The fit of the power law model could be better than the inverse-square model *if* the data does not actually have an inverse-square behavior.
5. A brighter bulb would shift increase all readings proportionately. So, *a* and *C* would increase, but *b* would stay the same.

Lesson 8-4

1. Hole at $x = -3$
2. Hole at $x = -1$
3. Hole at $x = -2$
4. Possible answer: The graph of a rational function has a hole for values of x that make both the numerator and the denominator equal to 0.
5. Hole at $x = 3$
6. Hole at $x = 1$
7. Hole at $x = \pm 1$

Lesson 8-7

1.

```
WINDOW
 ShadeRes=■
 Xmin=-5
 Xmax=20
 Xscl=2
 Ymin=-10
 Ymax=20
↓Yscl=2
```

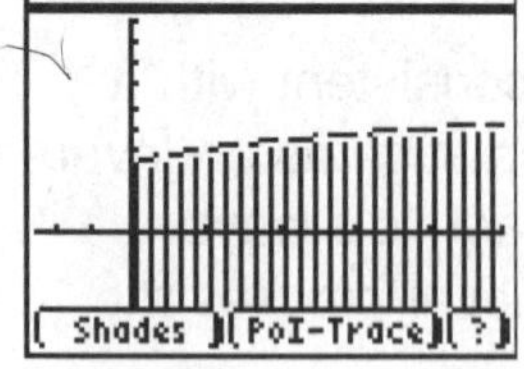

2.

```
WINDOW
 ShadeRes=3
 Xmin=0
 Xmax=40
 Xscl=5
 Ymin=-20
 Ymax=20
↓Yscl=5
```

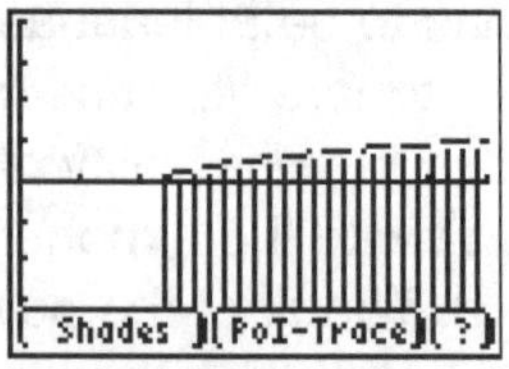

3.

```
WINDOW
 ShadeRes=3
 Xmin=-5
 Xmax=20
 Xscl=1
 Ymin=-10
 Ymax=10
↓Yscl=1
```

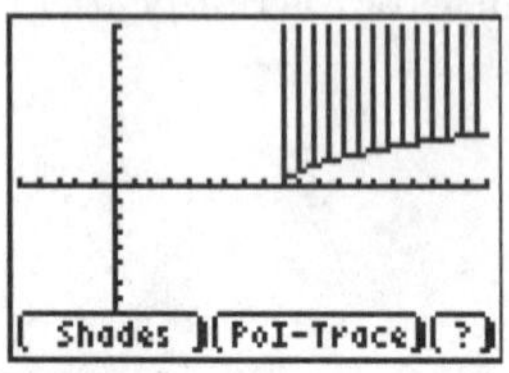

4.

```
WINDOW
 ShadeRes=3
 Xmin=0
 Xmax=20
 Xscl=5
 Ymin=-10
 Ymax=10
↓Yscl=2
```

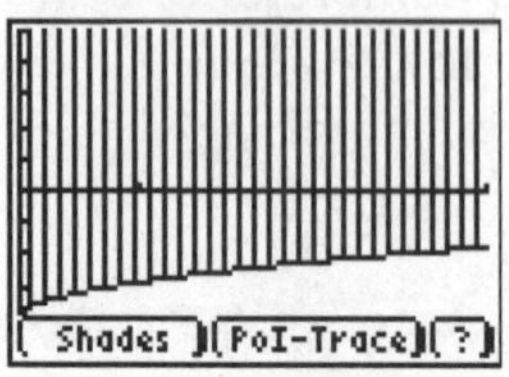

5.

```
WINDOW
 ShadeRes=3
 Xmin=0
 Xmax=20
 Xscl=5
 Ymin=-10
 Ymax=10
↓Yscl=2
```

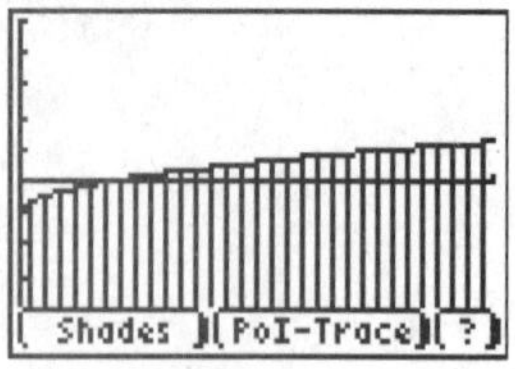

6.

```
WINDOW
 ShadeRes=3
 Xmin=0
 Xmax=10
 Xscl=1
 Ymin=-15
 Ymax=10
↓Yscl=2
```

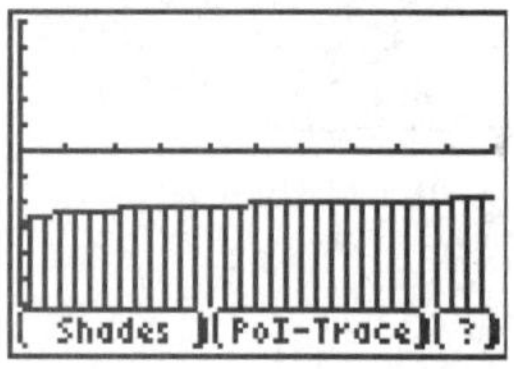

CHAPTER 9

Lesson 9-2

Activity 1

1. False
2. True
3. False
4. False
5. They are equal.

Activity 2

6. True
7. False
8. True
9. False
10. For $x > 1$, $x^2 > x > \sqrt{x}$. For $0 < x < 1$, $\sqrt{x} > x > x^2$.

Activity 3

11. 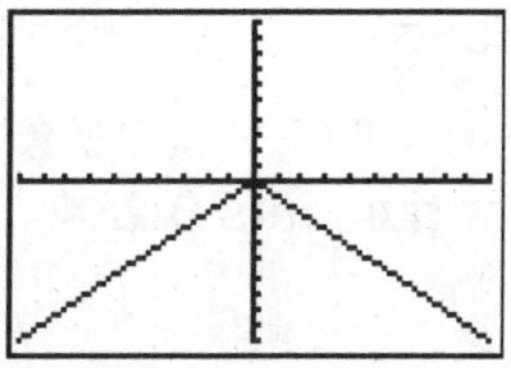

12. 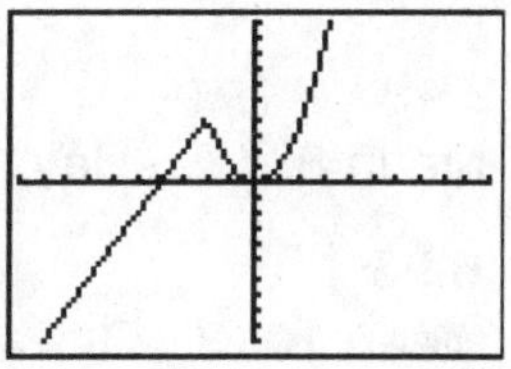

13. 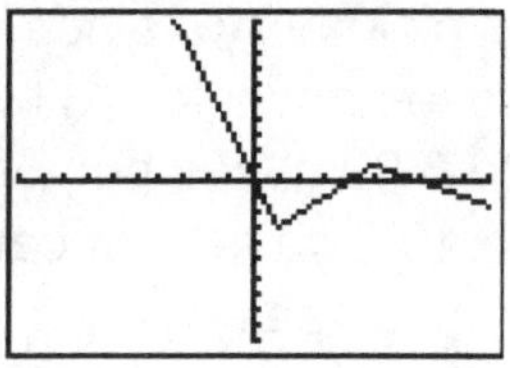

14. Possible answer: The logical test returns 1 if the statement is true and 0 if it is false. If the statement is false, the function is multiplied by 0 and all function values are 0. If the statement is true, the function is multiplied by 1 and all values stay the same.

15. Possible answer: The calculator gives the function a value of 0 when the *x*-values are not in the restricted domain. The connected mode would connect these values to the actual function values, creating extraneous lines.

Lesson 9-6 p. 43

1. 1st differences: 144, 240, 336, 432
   2nd differences: (constant): 96
   ratios: −6.2, ≈ 2.94, ≈ 1.92, ≈ 1.62

2. 1st differences: (constant): 1.17
   ratios: ≈ 1.89, ≈ 1.47, ≈ 1.32, ≈ 1.24, ≈ 1.20

3. 1st differences: 26.88, 107.52, 430.08, 1720.36
   2nd differences: 80.64, 322.56, 1290.28
   ratios: (constant): 4

4. 1st differences: 24, 42, 60, 78
   2nd differences: (constant): 18
   ratios: ≈ −1.67, 3.8, ≈ 2.05, ≈ 1.67

5. Quadratic; because quadratic functions have constant 2nd differences

Lesson 9-6 pp. 44–47

Typical plots

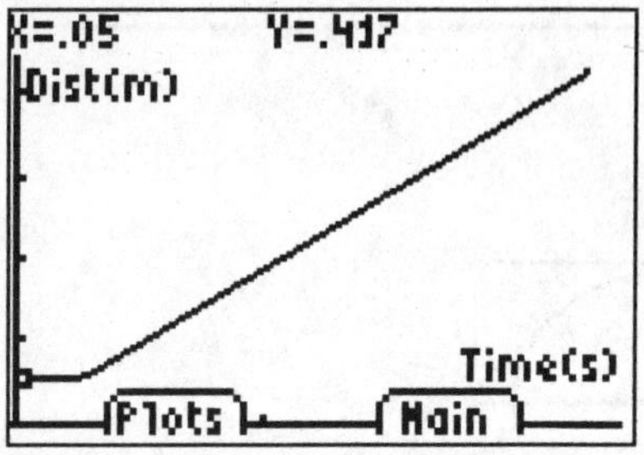

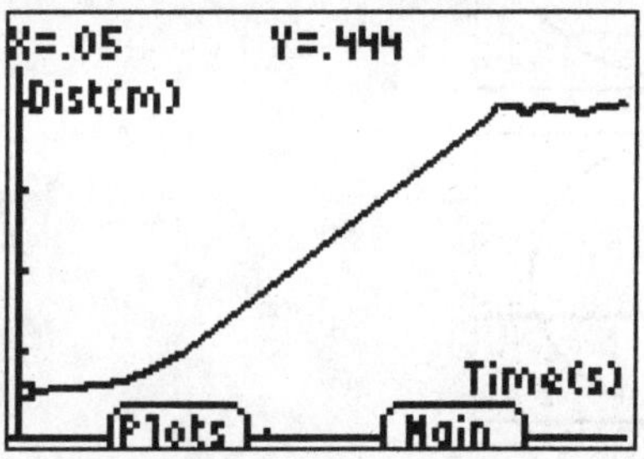

Try This

1. the third plot
2. time; seconds; distance of object from CBR 2™ motion detector; feet or meters
3. varies (should be half of a parabola, concave up)
4. a parabola (quadratic)
5. varies

Activity 2

1. varies (should be parabolic with increasing curvature)
2. 0° is flat (ball can't roll); 90° is the same as a free-falling (dropping) ball

Try This

3. Since the ball is at rest when released, *b* should approach zero for each trial.
   c should approach the initial distance, 0.5 meters (1.5 feet). *a* increases as the angle of inclination increases.

4. The ball's acceleration is due to the earth's gravity. So the more the ramp points down (the greater the angle of inclination), the greater the value of *a*. Maximum *a* occurs for $\theta = 90°$, minimum for $\theta = 0°$.

5. *a* is proportional to the sine of θ.

CHAPTER 10

Lesson 10-3

Activity 1

1. 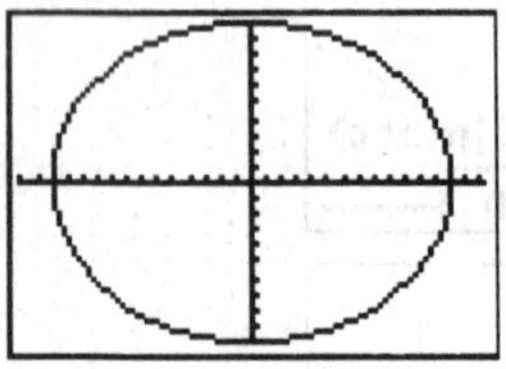

2. 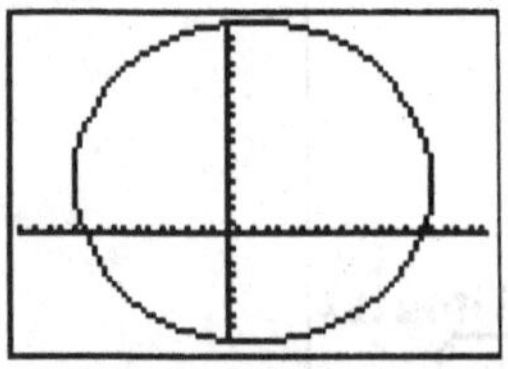

3. At the conics screen, press 2 instead of 1; because *b* is 10 and *a* is 13.

Activity 2

1. $\frac{x^2}{100} + \frac{y^2}{196} = 1$

2. 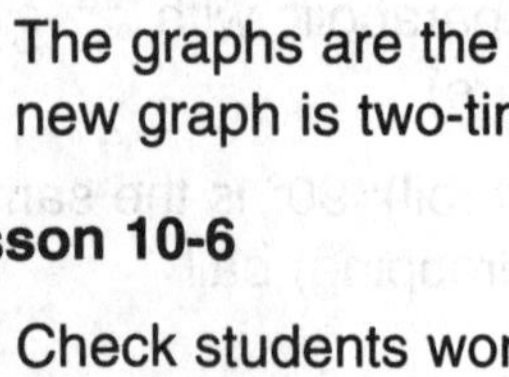 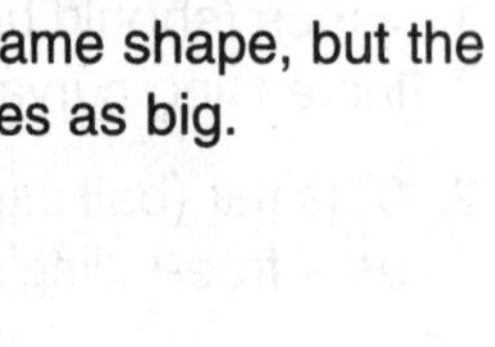

$\frac{(x)^2}{25} + \frac{(y)^2}{49} = 1$ $\frac{(x)^2}{100} + \frac{(y)^2}{196} = 1$

3. The graphs are the same shape, but the new graph is two-times as big.

Lesson 10-6

1. Check students work.

2. Check students work.

CHAPTER 11

Lesson 11-2

1. $\approx$ 16.9%

2. $\approx$ 2.6%

3. Possible answer: Use =INT(1000*RAND( )) 10 times. Consider numbers 0-284 successes.

Lesson 11-5

1. Mean 7.6 dev 2.49
Without outliers: Mean 8.3 Dev 1.41

2. Mean 12 Dev 6.96
Without outliers: Mean 10 .3 Dev 4.64

3. The Grocer should pick Jerry's since he produces 8.3 cases a week and his supply may drop to 6.89 cases per week. Mega can produce 10.3 cases but can also drop to only 5.66 cases per week.

CHAPTER 12

Lesson 12-2

1. −3, 3, 13, 27, 45, 67, 93, 123; 368

2. 0.25, 0.5, 1, 2, 4, 8, 16, 32; 63.75

3. 1.9, 2.2, 2.5, 2.8, 3.1, 3.4, 3.7, 4; 23.6

4. 20, 40, 60, 80, 100, 120, 140, 160; 720

5. −1, 4, 21, 56, 115, 204, 329, 496; 1224

6. 0.5, 2.5, 12.5, 62.5, 312.5, 1562.5, 7812.5, 39,062.5; 48,828

7. 110, 92, 74, 56, 38; 370

Lesson 12-4

1. 30

2. 1.9375

3. 88,573

CHAPTER 13

Lesson 13-3

1. $\sin 150° = 0.5$; $\cos 150° \approx -0.8660254$

2. $\sin 244° \approx -0.898794$; $\cos 244° \approx -0.4383711$

3. $\sin 90° = 1$; $\cos 90° = 0$

4. Possible answer: Use the TRACE feature to find the coordinates of the points where the circle crosses the axes. These points have coordinates (1, 0), (0, 1), (−1, 0), and (0, −1). The distance from each of these points to the origin is 1 unit. Therefore, the circle has a radius of 1 unit.

5. Possible answer: For an angle θ in standard position whose terminal side passes through point $P(x, y)$ on the unit circle, $\sin\theta = y$. The value of y is negative for points in Quadrants III or IV. Therefore, $\sin\theta$ is negative if the terminal side of θ lies in Quadrants III or IV.

Lesson 13-6

1. 82.05
2. 96.8
3. 37.8
4. 7.4 hours

CHAPTER 14

Lesson 14-3 p. 58

1. It is a possible identity.

2. $\sec x - \tan x \sin x = \frac{1}{\cos x} - \frac{\sin^2 x}{\cos x}$
$= \frac{1 - \sin^2 x}{\cos x}$
$= \frac{\cos^2 x}{\cos x = \cos x}$

3. It is a possible identity.

4. $\frac{1 + \tan x}{1 + \cos x} = \frac{1 + \frac{\sin x}{\cos x}}{1 + \frac{\cos x}{\sin x}}$
$= \frac{\frac{\cos x + \sin x}{\cos x}}{\frac{\sin x + \cos x}{\sin x}}$
$= \frac{\cos x + \sin x}{\cos x} \cdot \frac{\sin x}{\sin x + \cos x}$
$= \frac{\sin x}{\cos x} = \tan x$

Lesson 14-3 p. 59

1. $1 - \sin^2 \theta$; They are the same.

2. $\cos\theta$; They are the same.